T0406726

Palgrave Studies in Environmental Sustainability

Series Editor

Robert Brinkmann, Hofstra University, Hempstead, NY, USA

The series will take advantage of the growing interest in a number of environmental issues in sustainability, particularly in those that focus on interdisciplinary approaches to solving real-world problems. Unfortunately, many of the themes of sustainability address short- and long-term survivability of our planet and there is a pressing need for access to information to those who seek to solve sustainability problems. Thus, the scope of the book series seeks to be comprehensive within an environmental sustainability framework. The topics covered in the series range from climate change to public land management. While no series can ever be truly comprehensive, the series will provide one of the most definitive surveys of knowledge in the area of environmental sustainability.

Éloi Laurent

Toward Social-Ecological Well-Being

Rethinking Sustainability Economics for the 21st Century

Éloi Laurent
OFCE/Sciences Po,
Ponts Paris Tech
Stanford University
Paris, France

Palgrave Studies in Environmental Sustainability
ISBN 978-3-031-38988-7 ISBN 978-3-031-38989-4 (eBook)
https://doi.org/10.1007/978-3-031-38989-4

Cover credits: Isnurnfoto/Alamy Stock Photo

This Palgrave Macmillan imprint is published by the registered company Springer Nature Switzerland AG
The registered company address is: Gewerbestrasse 11, 6330 Cham, Switzerland

For Sylvie, Lila and Jonas, my holistic prosperity

Contents

1	**Prologue: From Economic to Holistic Sustainability**	1
	A Brief History of Economic (Un)Sustainability	2
	Social-Ecological Well-Being as the Cornerstone of Holistic Sustainability	9
	References	13
2	**Predicament: Our Intertwined Crises**	17
	The Planetary Health Crisis	17
	The Inequality Crisis	29
	Inequality Increases the Need for Environmentally Harmful and Socially Unnecessary Economic Growth	35
	Inequalities Increase the Ecological Irresponsibility of the Richest, Within Each Country and Between Nations	37
	Inequalities, Which Affect the Health of Individuals and Groups, Diminish the Social-ecological Resilience of Communities and Societies and Weaken Their Collective Ability to Adapt to Accelerating Global Environmental Change	37
	Inequalities Reduce the Political Acceptability of Environmental Concerns and the Ability to Offset the Potential Socially Regressive Effects of Environmental Policies	38
	The Cooperation Crisis	38
	References	51

3 Vision: Holistic Sustainability 55
Framework and Visualization: The Social-Ecological Loop 55
Social-Ecological Well-Being in Motion: Full Health, Just Transition, Social-Ecological Regeneration 61
Full Health 61
Just Transitions 76
Social-Ecological Regeneration 86
References 94

4 Policy: Sustainable Pathways 99
From Growth Policy to Well-Being Policies 99
From Cost–Benefit to Co-Benefits 111
From Digital Acceleration to Social Continuity 123
Mobilizing People and Activating States 130
References 134

5 Narrative: Reimagining Economics 137
The Doom Narrative 138
The Tech Narrative 138
The Economic Narrative 140
The Social-Ecological Well-Being Narrative 144
References 145

Index 147

List of Figures

Fig. 2.1	Breakdown of human development progress for OECD countries, 1870–2007 (*Source* Prados de la Escosura [2015] and author's calculations)	19
Fig. 3.1	**a–d**. Four social-ecological visions. **a**. Disconnected economics. **b**. Sustainable development. **c**. Planetary boundaries. **d**. Donut Economy	57
Fig. 3.2	Social-ecological crisis	59
Fig. 3.3	**a**. Social-ecological attrition. **b**. Social-ecological revival	60
Fig. 3.4	Full health: dimensions and indicators	71
Fig. 3.5	Percentage of total health expenditure devoted to prevention for 2009–2018, selected countries (*Source* Eurostat)	73
Fig. 3.6	Percentage of total health expenditure devoted to prevention in 2018, selected countries (*Source* Eurostat)	73
Fig. 3.7	Support for carbon taxation, without and with social compensation (% of respondents) [*Source* Dechezleprêtre et al. (2022b)]	84
Fig. 4.1	Number of countries that have launched well-being initiatives, 2000–2021 (*Source* OECD, own elaboration)	105
Fig. 4.2	Co-benefits and social savings (*Source* Own elaboration)	121

List of Tables

Table 2.1 Annual average growth rate, 1870–2007 (%) 18
Table 2.2 Health impacts of climate change 23
Table 3.1 Indicators and data of full health 72
Table 3.2 A typology of environmental inequality 81
Table 4.1 Three post-growth streams 104

CHAPTER 1

Prologue: From Economic to Holistic Sustainability

Sustainability theory and policy, as they stand, present an intriguing paradox, which is the starting point of this book: sustainability is currently understood in the sense of environmental sustainability yet it was born in the modern era as economic sustainability.

At its core, environmental sustainability means preserving the key elements of the human environment over time: a stable climate, healthy ecosystems, flourishing biodiversity. Economic sustainability, on the other hand, essentially means preserving the resources that allow for economic activity: minerals, energy, water, food, etc. and of course human labor. While the two notions are intertwined and even embedded and hierarchized (environmental sustainability is obviously the condition for economic sustainability), they are not equivalent in terms of analysis (sustainability of what and for whom?) and policy (how to achieve sustainability?).

Against this backdrop, sustainability analysis has become a science of consequences rather than causes. Sustainability scholars are hard at work documenting the tangible systemic crisis of our biosphere and with tremendous and growing precision, but the economic roots of this crisis are rarely exposed, examined or addressed. Environmental unsustainability is disconnected from economic sustainability.

É. Laurent, *Toward Social-Ecological Well-Being*, Palgrave Studies in Environmental Sustainability, https://doi.org/10.1007/978-3-031-38989-4_1

To turn this puzzling paradox into a fruitful beginning, one needs to uncover the economic origins of contemporary sustainability analysis at the turn of the nineteenth century. At that time, two towering figures of political economy would lay out the basic principles of our understanding of sustainability for the next two centuries.

A Brief History of Economic (Un)Sustainability

The seminal contribution in economic sustainability is the work of the Reverend Thomas Robert Malthus, who harbored a cynical view of humanity's fate in the face of its reproduction impetus and production limitations, a view he developed in his brief *Essay on the Principle of Population*, first published in 1798 (Malthus 1998).

Malthus had two questions in mind: What are "the causes that have hitherto impeded the progress of mankind towards happiness"? And what is "the probability of the total or partial removal of these causes in future"? As to the first question, he answers straightforwardly: "the constant tendency in all animated life to increase beyond the nourishment prepared for it".

The "Malthusian trap", as it came to be known (Clark 2007), is a mathematical conundrum: while population grows geometrically if left unchecked, food production only grows arithmetically. The result are inexorable famines which, in the eyes of Malthus, call for a structural solution: take the poorest part of the population out of the equation.

> *A man who is born into a world already possessed, if he cannot get subsistence from his parents on whom he has a just demand, and if the society do not want his labour, has no claim of right to the smallest portion of food, and, in fact, has no business to be where he is. At nature's mighty feast there is no vacant cover for him. She tells him to be gone, and will quickly execute her own orders, if he does not work upon the compassion of some of her guests.*
> (Thomas Malthus, An Essay on the Principle of Population 1798)

Because people tend to "produce" more people who cannot be sustained by adequate food supply, obstacles are needed to slow down population growth as soon as possible and force it to be constantly reduced to the level of the means of subsistence. Malthus knows only two ways to achieve this goal: "moral constraint" or "misery".

However debatable on empirical and ethical grounds, Malthus's reasoning can be seen as the first internally consistent sustainability analysis in the history of thought. It is based on a rudimentary model of human–environment interactions, grounded in social dynamics and revolves around the main variables of economic sustainability: natural resources, demography, income per capita, technical progress and social inequality. The key Malthusian assertion is eminently modern: demographic growth is incompatible with growth in living standards because of ecological constraints.

What is to be sustained in this view is human population, or rather a part of it, which is supposed to be the better part (Malthus harbors a form of plutocratic sustainability). Sustainability policy, in this brutal vision, relies on deeply questionable principles but it is clear that the solution to the sustainability problem is neither economic nor technical (Malthus does not mention the possibility of increasing the food supply or optimizing its distribution via price and/or market mechanisms). Rather, he sees the moral restriction of human excessive desires as a way out of the crisis he envisions, in line with Aristotelian philosophy (Laurent 2021).

While Malthus understands sustainability as an environmental problem with a cultural solution, with David Ricardo at the beginning of the nineteenth century, sustainability is translated into the formal language of political economy. Ricardo's vision had been preceded by that of von Carlowitz in the context of the German forestry industry, which is explicitly an economic approach to resource scarcity at a micro level (Spindler 2013).[1] In his book *Sylvicultura Oeconomica* (or the *Economic News and Instructions for the Natural Growing of Wild Trees*), published in 1713, Hans Carl von Carlowitz (1645–1714), having just been appointed as the head of the Saxon mining administration, coins the word *Nachhaltigkeit* which translates as "sustainable management principle of forestry" and implies that only as much wood as can be regrown should be cut. In a fascinating encounter between exploitation of non-renewable (mines) and renewable (forests) natural resources, sustainability was thus born as "continual, steady and sustained usage" of natural resources, its framework and

[1] This vision, as well as that of Ricardo, contradicts the environmentally naive Jean-Baptiste Say who writes in his *Treatise on Political Economy*: "Natural resources are inexhaustible, because otherwise we would not obtain them for free. Since they cannot be grown or exhausted, they are not a subject for economics" (Say 1840).

practices later imported into the US forestry industry by Gifford Pinchot under the term "conservationism" (Pinchot 1909).

Although grounded in political economy, von Carlowitz's insights stem from a criticism of economic rationale, namely the clearing of forests for cultivated fields (not unlike what happens in the Amazon Forest today), agriculture being the specific resource on which Ricardo, the inventor of economic sustainability, based his own understanding.

For Ricardo, unsustainability arises not as a moral issue but as the result of an economic problem: the law of diminishing returns. The most fertile lands are the first to be cultivated, but as the pressure of demand increases due to the increase in population, less fertile lands must in turn be cultivated and exploited. Because these new lands are less productive, their unit cost of production is higher, the selling price required for production on these new lands being also higher. The owners of the most fertile land thus benefit from a "differential" rent and "diminishing returns" appear because of the lower productivity of lands gradually cultivated under demographic pressure. This structural, natural decline in land productivity is such a powerful force that, according to Ricardo, it can drive the whole economy from a "progressive state" to a "steady state", a gloomy prospect he explicitly considers at the very end of Chapter V of his *Principles of Political Economy* (Ricardo 1999).

With Ricardo, economic activity becomes a proxy for human prosperity and exploitation; rent and profit, supposed to be the key drivers of this prosperity, take center stage in sustainability analysis. In the very first paragraphs of the Preface to the third and final edition of the *Principles*, Ricardo writes:

> *The produce of the earth—all that is derived from its surface by the united application of labour, machinery, and capital, is divided among three classes of the community; namely, the proprietor of the land, the owner of the stock or capital necessary for its cultivation, and the labourers by whose industry it is cultivated. But in different stages of society, the proportions of the whole produce of the earth which will be allotted to each of these classes, under the names of rent, profit, and wages, will be essentially different; depending mainly on the actual fertility of the soil, on the accumulation of capital and population, and on the skill, ingenuity, and instruments employed in agriculture. To determine the laws which regulate this distribution, is the principal problem in Political Economy.*

Whenever profits fall, the incentives for capital accumulation diminish. At a certain general rate of profit, which has become very low, the accumulation of capital comes to a complete halt. Indeed, at this level, the capitalist class (farmers, manufacturers) can no longer be compensated for their risk taking. Accumulation of capital stops, which also brings to a standstill the demand for labor and the population. In a very peculiar way, the steady-state system is stable and possibly sustainable if there is no further increase in population.

But this is not the end of the Ricardian story: international exchange (the premises of contemporary globalization) and technological progress (in agriculture and then industry) are powerful solutions to the law of diminishing returns. But for Ricardo international exchange only provides a short-term solution, as the fertility of the trading partners' land is also declining (the inevitable depletion of good land can only be delayed, by international trade). The only genuinely sustainable solution is thus technical progress feeding labor productivity increasing income per capita as the population expands, in other words, perpetual economic growth.

Under the influence of Ricardo, in the early nineteenth century sustainability thus became an economic issue with an economic solution. Ricardo effectively paved the way for the adoption of technical progress-driven economic growth as the central variable of economic sustainability. With the invention of the concept of gross domestic product in 1934 by Simon Kuznets, GDP growth becomes the state variable[2] of economic sustainability, turning the Ricardian intuition into actual policy.

For a brief moment, at the end of the 1960s, Ricardian economic sustainability was contested by Paul Ehrlich, who attempted to bring back the "Malthusian trap" argument in his blockbuster book (Ehrlich 1968). "Nothing could be more misleading to our children than our present affluent society", he writes, predicting that "hundreds of millions of people will starve to death in the 1970s and 1980s" and recommending "population control". Just as Malthus did not see the rise of exponential economic growth around him, Ehrlich does not perceive that the "bomb" he warned was about to explode was being defused by the fall in fertility rates in developing countries (the annual growth rate of the world's population reached its peak around 2% in the mid-1960s, at the exact moment of the publication of Ehrlich's book, only to begin an

[2] A state variable is a given variable at time t that describes the state of a dynamic model.

immediate decline and to be almost halved since then).[3] This decline in fertility rates was accompanied by a typical Ricardian response, namely the "green revolution" in agriculture and the take-off of half a century of gradually liberalized globalization, the end of which we may finally be witnessing in the early 2020s.

The Malthusian prophecy was once again turned upside down: population growth became arithmetic while economic growth became exponential.

But just as the Ehrlich counter-attack was being parried and dismissed, the most consistent criticism of economic sustainability was released in the public debate with the "Limits to Growth" report, first published in 1972 (Meadows et al. 1972). The key argument of the team working on modelling simulations under the supervision of Dennis and Donella Meadows at MIT was simple: economic growth had become the driving factor of environmental unsustainability, threatening human well-being and, in the end, economic sustainability itself. This vision, echoing that of John Stuart Mill[4] at the peak of the industrial revolution, had two versions: a collapse by resource scarcity and a collapse by pollution.

While visionary politicians reacted immediately to these striking findings,[5] economists were quick to dismiss the scarcity scenario on the grounds that the model behind it relied on volumes and not prices. In doing so, economists overlooked the fundamental novelty of the report with respect to sustainability analysis: the key problem was not scarcity

[3] The latest projections by the United Nations (2022) suggest that in 2020, the growth rate of the global population fell below 1 per cent per year for the first time since 1950. Yet, the world's population is not projected to reach a peak (at around 10 billion people) before the 2080s.

[4] John Stuart Mill, who, in Chapter VI of Book IV of his *Principles of Political Economy*, published in 1848, devotes developments of astonishing modernity to what he calls, following Ricardo, the "stationary state".

[5] In a bold and visionary letter dated 14 February 1972 and addressed to the then President of the European Commission Franco-Maria Malfatti, the European Commissioner for Agriculture Sicco Mansholt, alarmed by the scope of the environmental crises revealed by the MIT 'Limits to Growth' report (Meadows et al. 1972), wrote: 'It is clear that tomorrow's society cannot be concentrated on growth, at least not as far as material goods are concerned' (Mansholt 1972). He went on to argue that the ten Member States of the European Economic Community (EEC) should stop directing their 'economic system to the search for maximum growth and to constant increase in the gross national product'. 'We would do well', he added, 'to examine how we could help in establishing an economic system which is no longer based on maximum growth per inhabitant' (ibid.).

but over-abundance, there were not too few resources at the disposal of humanity but far too many.

A new vision of sustainability, comprising the full set of social, economic and ecological parameters that it needed to be accurate, had emerged.

Conventional economics soon reiterated economic sustainability arguments, updating Ricardo to debunk Meadows. Drawing inspiration from Hotelling's work on the exploitation of exhaustible resources, John Hartwick, for instance, attempted to demonstrate that if revenues from the exploitation of natural resources were fully reinvested in the formation of productive capital, the total production capacity bequeathed to future generations would remain intact (the so-called "Hartwick rule"; Hartwick 1977; Laurent 2021). This proposal by Hartwick paved the way for the promotion of a "weak" form of sustainability that posited the easy substitution of natural capital by physical capital to maintain a consistent total capital stock over time.[6] This vision was expanded by Solow (1993), who popularized a view of weak sustainability whereby humanity was called on to preserve a "generalized capacity to produce economic well-being" in time. "The duty imposed by sustainability" Solow notes "is to endow [future generations] with whatever it takes to achieve a standard of living at least as good as our own and to look after the next generation similarly. We are not to consume humanity's capital, in the broadest sense."

This renewed stream of economic sustainability nurtured the approach developed in the Brundtland Report[7] (WCED 1987), which coined and popularized the notion of "sustainable development". Although the report attempted to link the issue of sustainability with that of inequality, it ended up recommending a piecemeal approach to sustainability, breaking it down into three dimensions (social, economic, environmental) and thus acknowledging the possibility of sustainability trade-offs in the spirit of the weak sustainability view (economic against social, social against environmental, and most of all, economic against environmental).

[6] Conversely, proponents of the "strong" form of sustainability argue that, all natural capital loss being irreversible, it should be kept whole.

[7] The Brundtland Report was named after the Commission's chair, Gro Harlem Brundtland. The notion of sustainable development was first introduced by the International Union for the Conservation of Nature's 1980 World Conservation Strategy titled "Living resource conservation for sustainable development".

Yet, economic sustainability proponents failed to provide a robust answer to the second set of scenarios in the "Limits to Growth" report simulations, which linked collapse not to resource scarcity but to pollution. The accelerating climate crisis (IPCC 2021) has become the obvious incarnation of these scenarios and, 50 years down the road, the undisputable validation of the Meadows team's intuitions.

By all accounts, in the early 2020s, economic sustainability has hit the reality wall of global environmental change. The "great acceleration" driven by economic growth has led to the trespassing of most "planetary boundaries" and ultimately "tipping points", pushing the biosphere toward near collapse. In 1944, at the time when the Bretton Woods conference set GDP as the gold standard of economic development for all the countries in the world, global GDP was $8 trillion (double what it was in 1900). It went on to reach $30,000 billion in 1975, rising to $60,000 billion in the year 2000 and exceeding $100,000 billion in 2015. With each threshold crossed, ecological damage has exploded. Economic growth, supposed to be the main driver of sustainability, has become the epicenter of unsustainability.[8]

While mainstream economists continue to advocate for ever-increasing growth, louder and louder voices in the environmental community denounce this destructive obsession. Indeed, the criticism of GDP and growth as intangible economic horizons at a time when ecological crises are accelerating has never been stronger and has taken center stage in major environmental international institutions (EEA 2021; IPCC 2021; Pörtner et al. 2021; IPBES 2022). Summing up those concerns, the 11,000 climate researchers assembled in the Alliance of World Scientists state in unprecedentedly clear-cut terms that, to meet the Paris targets and move toward net-zero societies, "economic growth must be quickly curtailed" to "maintain long-term sustainability of the biosphere", and that the goals of economic and other policymaking "need to shift from [gross domestic product (GDP)] growth … toward sustaining ecosystems and improving human well-being by prioritizing basic needs and reducing inequality" (Ripple et al. 2020).

[8] The damage to the biosphere before 1944 was negligible compared to that which occurred afterwards, starting with GHG emissions, the destruction of non-human species and the large-scale destruction of planetary ecosystems (seas and oceans, tropical forests, etc.). Cumulative CO_2 emissions amounted to 200 billion tons before 1944, while they are 1300 billion today (15% against 85% of the total).

By the same token, the Sixth Assessment Report by the IPCC (IPCC 2023) showcases the only viable climate scenario for humanity in the twenty-first century (a scenario dubbed "SSP1-1.9", or "Shared Socioeconomic Pathway 1–1.9", which foresees a stabilization of global warming at 1.6 degrees between 2041 and 2060 before witnessing a decrease to 1.4 degrees at the end of the twenty-first century), based on the work of Riahi et al. (2017) who have defined the SSP1 scenario in the following terms:

> *Sustainability – Taking the Green Road (Low challenges to mitigation and adaptation) The world shifts gradually, but pervasively, toward a more sustainable path, emphasizing more inclusive development that respects perceived environmental boundaries. Management of the global commons slowly improves, educational and health investments accelerate the demographic transition, and the emphasis on economic growth shifts toward a broader emphasis on human well-being. Driven by an increasing commitment to achieving development goals, inequality is reduced both across and within countries. Consumption is oriented toward low material growth and lower resource and energy intensity.*

If SSP1 is correctly understood as a possible path for humanity away from climate chaos, then it clearly translates into two important challenges: prioritizing well-being instead of GDP growth (again, see Laurent 2021) and reducing inequality both between and within countries. Thus emerges the need for a new vision of sustainability, holistic rather than economic.

Social-Ecological Well-Being as the Cornerstone of Holistic Sustainability

What can we learn from this brief overview to overcome the dead end of economic (un)sustainability? First and foremost, that sustainability cannot be reduced to economic sustainability: we are heading for environmental collapse with GDP as a compass and we need to think again about the fundamental questions of sustainability: sustainability of what and for whom? With what means for what ends? Second, piecemeal sustainability is an illusion: we must develop an holistic vision of sustainability linking the different dimensions of dynamic human well-being in a consistent framework rather than relying on short-term sustainability trade-offs.

In trying to provide answers to these two questions, this book's central contribution is to argue that what should be sustained is not economic activity but *social-ecological well-being defined as a combination of planetary health, cooperation and justice resulting in human holistic prosperity*.

The discipline of economics can be of great help in articulating these essential dimensions of human existence, provided it is deployed in all its richness and complexity. The history of thought, too often forgotten in existing conventional economics textbooks, is not just about remembering ideas and intuitions from the past. Above all, it is about confronting yesterday's reflections with today's challenges in order to understand how the problems and possible solutions can look alike and how they differ from one era to another. In this respect, numerous streams of economic reasoning, old and new, will irrigate the reflections presented in this book (Box 1.1).

Box 1.1: A Brief Overview of Contemporary Eco-Diversity

Ecological economics

Key idea: the economy is a subsystem of society, which is itself a subsystem of the biosphere.

Feminist economics

Key idea: informal, non-market activities, in particular care and household chores/tasks, which are mainly carried out by women, form the bedrock of economic systems.

Institutional economics

Key idea: Institutions (defined as formal and informal rules), rather than markets, govern the most important decisions of economic actors.

Economics of complexity

Key idea: the economy is a complex system, interacting with other complex systems (such as the environment).

Evolutionary economy

Key idea: Economic systems are not static and change over time under the influence of technological innovation and economic cycles.

Behavioral economics

Key idea: the idea of a *Homo economicus* whose decisions would be rational is not robust.

Marxian political economy

Key idea: power asymmetries and the unequal distribution of resources between actors are at the heart of economic dynamics.

Source Adapted from Brand-Correa et al. (2022).

The first pillar of social-ecological well-being is *planetary health*. As the COVID-19 pandemic has made so clear, health is a collective affair. The more public health is, the healthier individuals and groups are; the more individuals are left alone to deal with their health, the more health deteriorates for everyone. Yet, contrary to these evidence-based principles, a separatist approach has been deployed in three ways in past decades: the priority given to individual behaviors over public health; the usefulness in this perspective of digital tools for health individualization; and the detachment of health issues from ecological challenges.

Contemporary health studies show, on the contrary, that social logics are the key to understanding salutogenesis (the factors that foster health) as well as pathogenesis (the factors that spread disease) of populations. This is shown, for example, in the Marmot report on health equity in the United Kingdom published in 2020, whose authors are rightly alarmed at the drop in life expectancy in the most deprived communities in the country after decades of progress, due to the austerity imposed on healthcare services and social disintegration (Marmot 2020).

This collective nature of health, which is so clear when considering the importance of social ties in the evolution of life expectancy,[9] must be extended to the biosphere: there is no human health without the health of ecosystems and the biodiversity that sustain them. Hence the concept of planetary health, defined as "the health of human civilization and the state of the natural systems on which it depends" (Horton et al. 2015).

Health is fundamentally connected to environmental conditions of living but, contrary to Malthus and Ricardo's understanding of population dynamics, the key issue here is not just survival of humanity through nutrition but fostering long, healthy and whole lives entailing physiological and psychological well-being, stable climate, ecosystems, biodiversity and yes, nutritious and healthy food. Health is in fact the key mediation between humans and the rest of nature but also the condition of economic activity, which is only a facet of social cooperation.

[9] The most astonishing, robust and valuable lesson of the uniquely insightful Harvard Study of Adult Development (HSAD) is simple: social relationships are the key to mental and psychological health and the most powerful determinant of life expectancy and happiness (Waldinger and Schulz 2023).

Cooperation is indeed the endless source of social innovation which can renew, reinvent and sustain human organizations and direct them to achieve and maintain human well-being. Social innovation goes beyond labor productivity enhancing technical progress, as understood by Ricardo, and includes such institutions as labor contract, rule of law or modern universities that can foster social cooperation over long periods of time, sometimes centuries. Because we humans form a cooperative species, we rely on social innovation to achieve our goals, especially new ones such as mitigating and adapting to environmental crises like climate destabilization.

In this respect a new set of questions have arisen with the digital transition that has emerged since the early 1990s. The digital revolution is technical progress based on allegedly de-materialized human intelligence acceleration that was supposed to be a viable sustainability strategy accelerating cooperation without ecological cost. But digital innovation has in fact increased unsustainability via re-materialization and de-socialization. The possibility of a conflict between digital collaboration and social cooperation is thus a key contemporary issue to be explored, with multiple ramifications.

Finally, cooperation and health fundamentally rely on *justice*, which is both a goal and a method for human communities. Malthus thought that injustice was the solution to unsustainability but he was deeply mistaken. The current ecological crises are also social crises: the sumptuous riches of our planet are being squandered for the benefit of a handful among us, mostly in already immensely wealthy countries, and our societies have become increasingly unequal, fragmented and polarized over the past 40 years, while environmental degradation has accelerated to reach new levels. The inequality crisis and the ecological crises go hand in hand. The 35 countries considered to be the richest, which represent only 15% of the world's population, are together responsible for 75% of the disproportionate consumption of natural resources since 1970. And half of the CO_2 emissions since 1990 are the result of only 10% of humans. In order to break this inequality-unsustainability nexus, social justice should foster the social cooperation needed to mitigate our ecological crises. We must find practical ways to reverse the vicious social-ecological spiral in which we are caught (we destroy the habitat that contains us) to enter a virtuous circle where ecological interdependence and social cooperation feed each other instead of devouring one another.

The long-term prosperity of humanity does not rely on our collective ability to generate material wealth in perpetuity (long-term economic growth depending on population and technical progress). It relies on generating health and fostering cooperation informed by justice. This is what we should sustain. Using social-ecological well-being to rethink sustainability economics for the twenty-first century means recognizing that justice sustains cooperation which underpins health that connects us to the biosphere.

Within this framework, this book attempts first to explain why the three key dimensions of sustainability are jointly in crisis (Chapter 2), what vision can articulate those dimensions to rethink sustainability economics for our century (Chapter 3) and what practical policies should be undertaken to give life to these visions (Chapter 4), before concluding on the need to reinvent the narratives that sustain economic analysis (Chapter 5).

References

Brand-Correa, L., et al. (2022), "Economics for People and Planet: Moving Beyond the Neoclassical Paradigm", *The Lancet Planetary Health*, vol. 6, no. 4, pp. e371–e379.

Clark, G. (2007), *A Farewell to Alms: A Brief Economic History of the World*. Princeton: Princeton University Press.

EEA (2021), Growth Without Economic Growth. Briefing No. 28/2020. https://doi.org/10.2800/492717.

Ehrlich, P.R. (1968), *The Population Bomb*. Reprint of the 1968 ed. published by Ballantine Books, New York, in series: A Sierra Club-Ballantine.

Hartwick, J.M. (1977), "Intergenerational Equity and the Investing of Rents from Exhaustible Resources", *The American Economic Review* (Dec).

Horton, R., et al. (2015), "Planetary Health: A New Science for Exceptional Action", *The Lancet*, vol. 386, no. 10007, pp. 1921–1922.

IPBES (2022), "Summary for Policymakers of the Methodological Assessment of the Diverse Values and Valuation of Nature of the Intergovernmental Science-Policy Platform on Biodiversity and Ecosystem Services". In: U. Pascual, P. Balvanera, M. Christie, B. Baptiste, D. González-Jiménez, C.B. Anderson, S. Athayde, R. Chaplin-Kramer, S. Jacobs, E. Kelemen, R. Kumar, E. Lazos, A. Martin, T.H. Mwampamba, B. Nakangu, P. O'Farrell, C.M. Raymond, S.M. Subramanian, M. Termansen, M. Van Noordwijk, and A. Vatn (eds). IPBES secretariat, Bonn, Germany, 37 pages. https://doi.org/10.5281/zenodo.6522392

IPCC (2021), "Climate Change 2021: The Physical Science Basis. Contribution of Working Group I to the Sixth Assessment Report of the Intergovernmental

Panel on Climate Change". In: V. Masson-Delmotte, P. Zhai, A. Pirani, S.L. Connors, C. Péan, S. Berger, N. Caud, Y. Chen, L. Goldfarb, M.I. Gomis, M. Huang, K. Leitzell, E. Lonnoy, J.B.R. Matthews, T.K. Maycock, T. Waterfield, O. Yelekçi, R. Yu, and B. Zhou (eds). Cambridge University Press.

IPCC (2023), "Summary for Policymakers. In: Climate Change 2023: Synthesis Report. Contribution of Working Groups I, II and III to the Sixth Assessment Report of the Intergovernmental Panel on Climate Change" [Core Writing Team, H. Lee and J. Romero (eds.)]. IPCC, Geneva, Switzerland, pp. 1–34. https://doi.org/10.59327/IPCC/AR6-9789291691647.001

Laurent, É. (2021), "Introduction to Economics and Sustainability". In: *The Palgrave Handbook of Global Sustainability*. Cham: Palgrave Macmillan. https://doi.org/10.1007/978-3-030-38948-2_98-1.

Malthus, T. (1798[1998]), *An Essay on the Principle of Population, as It Affects the Future Improvement of Society*, London, Printed for J. Johnson, in St. Paul's Church-Yard, Electronic Scholarly Publishing Project. http://www.esp.org/books/malthus/population/malthus.pdf.

Mansholt, S. (1972), "Letter to the President of the European Commission Franco-Maria Malfatti", The Mansholt Letter, Rotterdam, Het Nieuwe Instituut. https://themansholtletter.hetnieuweinstituut.nl/sites/default/files/brief_mansholt_malfatti_en1.pdf.

Marmot, M., Allen, J., Boyce, T., Goldblatt, P., and Morrison J. (2020), Health Equity in England: The Marmot Review 10 Years On. Institute of Health Equity. https://health.org.uk/publications/reports/the-marmot-review-10-years-on.

Meadows, D., Meadows, D.L., Randers, J., and Behrens W.W. (1972), *The Limits to Growth; A Report for the Club of Rome's Project on the Predicament of Mankind*, New York, Universe Books.

Pinchot, G. (1909), "Conservation". In: *Addresses and Proceedings of the First National Conservation Congress*, August 26–28, Seattle. Washington: The executive committee of the National Conservation Congress, p. 72.

Pörtner, H.O., et al. (2021), IPBES-IPCC co-Sponsored Workshop Report on Biodiversity and Climate Change; IPBES and IPCC.

Riahi, K., et al. (2017), "The Shared Socioeconomic Pathways and Their Energy, Land Use, and Greenhouse Gas Emissions Implications: An Overview", *Global Environmental Change*, vol. 42, pp. 153–168.

Ricardo, D. (1999 [1821]), *On the Principles of Political Economy and Taxation*. London: John Murray, Albemarle-Street. Library of Economics and Liberty, 1999. https://www.econlib.org/library/Ricardo/ricP.html.

Ripple, W.J., Wolf, C., Newsome, T.M., Barnard, P., and Moomaw, W.R. (2020), "World Scientists' Warning of a Climate Emergency", *BioScience* vol. 70, no. 1, pp. 8–12. https://doi.org/10.1093/biosci/biz088/5610806.

Say, J.-B. (1840), *Cours complet d'économie politique pratique*. Paris: Guillaumin.

Solow, R. (1993), "An Almost Practical Step Toward Sustainability", *Resources Policy*, Elsevier, vol. 19, no. 3, pp. 162–172, September.

Spindler, E. (2013), "The History of Sustainability: The Origins and Effects of a Popular Concept". In: I. Jenkins, and R. Schröder (eds), *Sustainability in Tourism*. Wiesbaden: Springer Gabler. https://doi.org/10.1007/978-3-8349-7043-5_1.

Waldinger, R., and Schulz, M. (2023), *The Good Life*. New York: Simon and Schuster.

World Commission on Environment and Development (WCED) (1987), *Our Common Future*. Oxford and New York: Oxford University Press.

CHAPTER 2

Predicament: Our Intertwined Crises

To start developing a new vision for sustainability, one has to begin with the world as it is and to review the different facets of the current unsustainability crisis, which is original in many ways. While doing so, I will start building bridges between the different dimensions of sustainability as I have defined it in the introduction.

The Planetary Health Crisis

There is hardly any better indicator of the extraordinary human development progress over the last century than health, starting with life expectancy, which has increased by 40 years, or almost two generations, since 1900.[1] In the course of this unprecedented progress, medical and technical innovations that made it possible to decipher, contain and push back infectious diseases in the first half of the twentieth century have played a major role. But this role would have been marginal without the democratization of the quality of life through the expansion of labor

[1] Life expectancy first increased among the youngest (in the first half of the twentieth century) before progressing among the oldest (in the second half, especially the last quarter, of the twentieth century and the first quarter of the twenty-first century).

É. Laurent, *Toward Social-Ecological Well-Being*, Palgrave Studies in Environmental Sustainability, https://doi.org/10.1007/978-3-031-38989-4_2

rights and social protections, progress in public education and access to care, and the mainstreaming of better hygiene and healthy nutrition.

To put it simply, life expectancy has been the most powerful driver of human development in the twentieth century, and the increase in life expectancy in the twentieth century has been the most significant event in human evolution. Data compiled by Prados de la Escosura (2015) suggest that, for all countries of the world, human development made considerable progress between 1870 and 2007, its average level rising by a factor of six. But these data also show that improvement in health was the most important driver in the human development index in the past 140 years, both for OECD and European countries and for the rest of the world (Table 2.1 and Fig. 2.1).

Yet, continued increase in life expectancy is a very recent phenomenon in the history of humanity and can be reversed: between the sixteenth century and the nineteenth century, life expectancy in the United Kingdom fluctuated under the impacts of various shocks (wars, diseases, etc.), oscillating between 30 and 40 years (Riley 2005); life expectancy declined in the United States between 2014 and 2017 (even before the Covid pandemic hit), an unprecedented regression since the Second World War, and even more dramatically in 2020 and 2021 due to Covid.

Human health depends on social bonds and natural ties: social bonds give us life, preserve and maintain it, while natural ties sustain it. There is indeed a vital interconnection between what we refer to as our "environment" and our health. Scholars have long highlighted the positive impact that protecting the environment can have on people's health and well-being, and conversely have precisely documented the negative health impact that environmental crises are already having.

Table 2.1 Annual average growth rate, 1870–2007 (%)

	Human development index	*Contribution of life expectancy*	*Contribution of education*	*Contribution of GDP per capita*
World	1.3	0.6	0.5	0.2
OECD	1.1	0.5	0.4	0.2
Non-OECD countries	1.7	0.7	0.8	0.2

Source Prados de la Escosura (2015) and author's calculations

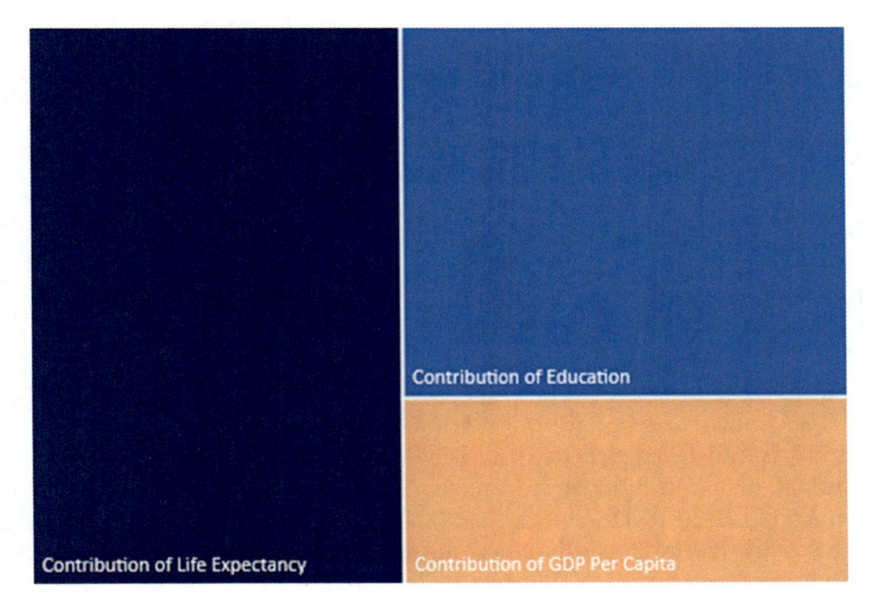

Fig. 2.1 Breakdown of human development progress for OECD countries, 1870–2007 (*Source* Prados de la Escosura [2015] and author's calculations)

From the early industrial revolution in Europe, public authorities understood the importance of preserving human health from the toxic flows of various types of pollution generated by the outburst of economic activity.[2] But the contemporary crisis of planetary health is such that the lives of tens of millions of people across the planet are now at stake each year.

Several studies published in quick succession at the start of 2022 attest to this new reality. The first series of studies relate to the origins of the COVID-19 epidemic (Worobey et al. 2022; Pekar et al. 2022) and the assessment of its impact for humans (Haidong et al. 2022). Against the backdrop of countless conspiracy theories, a scientific consensus has recently solidified as to the emergence of SARS-CoV-2 at the origin of the COVID-19 pandemic: this pandemic results "very probably" from a

[2] The first laws regulating French industrial establishments and in particular the imperial decree of October 15, 1810 was the first legislation in the world regulating pollution (it was extended by the law of December 19, 1917).

virus carried by an animal (probably the horseshoe bat) transmitted to the human species (as was the case with SARS-CoV-1 in 2002–2003) with or without the help of an intermediary host and it is very likely on the live animal market in the Chinese city of Wuhan that this transmission took place for the first time. As underlined by a related study on the proven link between the multiplication of zoonoses[3] and the destruction of natural environments by human activities, it is therefore clear that human health is "inextricably linked" to the health of ecosystems and biodiversity (Lawler et al. 2021). While 2022 has seen the emergence of the most serious episode of avian flu in history, deforestation, the destruction of biodiversity and intensive agriculture are more than ever appearing as direct threats to human health.

The human cost of the COVID-19 pandemic for the 2020–2021 period is much greater than initially estimated: the estimated excess mortality due to the global pandemic in 191 countries and territories from January 1, 2020 to December 31, 2021 shows a difference of 1–3 with the official figures: taking into account errors and mistakes in the census of deaths from COVID-19, we should deplore not 5,940,000 deaths worldwide over this period, but 18,200,000 (a global excess mortality of around 16%). For some countries, such as India, the difference is considerable: from 489,000 official deaths to 4,070,000 estimated deaths. For France, this same difference is substantial: from 122,000 to 155,000, i.e., a difference equivalent to the official deaths of the first wave in spring 2020. And yet this global estimate is based on the figure of 17,900 Chinese deaths (nearly four times more than officially announced), which is very difficult to believe (Haidong et al. 2022). At this point, it is possible that the Covid pandemic has killed about 25 million humans since its emergence in 2020. As for the psychological impact of the pandemic, the WHO estimates that the prevalence of symptoms of anxiety and depression increased by 25% for the year 2020, but here too, the real toll is probably much heavier.[4]

[3] A zoonosis is an infectious disease that can be transmitted from a non-human animal to humans.

[4] "COVID-19 pandemic triggers 25% increase in prevalence of anxiety and depression worldwide", WHO, March 2022, https://www.who.int/news/item/02-03-2022-covid-19-pandemic-triggers-25-increase-in-prevalence-of-anxiety-and-depression-worldwide.

But the threat of biodiversity commodification and ecosystems degradation to human well-being goes way further than the multiplication of zoonotic events.

As biodiversity underpins the vitality of ecosystems which sustain human communities, the destruction of biodiversity leads to the degradation of ecosystems—which are also subject to direct attacks (fires, deforestation, etc.)—and eventually the degradation of human life. Thus, according to the IPBES (IPBES 2019), the contributions of nature to the quality of human life have been declining since 1970 for fourteen of the eighteen categories studied (i.e. 77%). For some countries like Australia, which for decades have centered their development strategy on growth fueled by extraction, the awakening is brutal: the five-year report on the natural state of the country published in spring 2022 shows a general degradation of ecosystems, nineteen of which are on the verge of collapse, while the alteration of biodiversity, unparalleled in the world, is labelled "shocking" by experts.[5]

This alteration of the biosphere by humanity is obviously extremely costly for human development: biodiversity supports ecosystem functions, which themselves support nature's contributions and benefits, which themselves support human well-being. The destruction of the biosphere is therefore an irrational self-destruction—an act of arson in the web of life.

The climate crisis similarly highlights the intersection of health and environmental themes: the second part of the IPCC's AR6 report (IPCC 2022), the 3676 pages of which focus on impacts, adaptation and vulnerability, contains no less than 4853 occurrences of the word "health". The experts who developed the "planetary health"[6] approach drew up a precise inventory of the impact of the ongoing climate crisis (Table 2.2), documenting direct effects on health (heat waves, floods, droughts, hurricanes and storms aggravating global morbidity and mortality) as well as indirect effects (airborne diseases, food insecurity and malnutrition,

[5] The 2021 State of the Environment report, accessible at soe.dcceew.gov.au.

[6] In 2015, the Rockefeller Foundation-Lancet Commission on Planetary Health published its report "Safeguarding human health in the Anthropocene epoch" in *The Lancet*, defining "planetary health" as a solutions-oriented, transdisciplinary field and social movement focused on analyzing and addressing the impacts of human disruptions to Earth's natural systems on human health and all life on Earth (Redvers 2021).

population displacements, and diseases and mental injuries resulting from meteorological events, extremes and global warming).[7]

The projected impacts of climate change are expected to intensify in the coming years and decades, as highlighted in *The Lancet* Countdown 2020 report (Watts et al. 2021), which reported the bleakest outlook since its inception in 2015. The report also revealed various climate-related health effects in all the world's countries in multiple forms: heat waves leading to thousands of additional deaths annually, particularly among the elderly and those with chronic diseases; changing environmental and ecological conditions making some areas more vulnerable to various infectious diseases; indirect health impacts linked to occupational health; and stresses that affect mental health and well-being, etc.

As dire as this picture may already be, it is still incomplete: other studies focus on the health consequences of the "normal" functioning of the global economic system and aim to estimate the health impact of pollution, mainly air and water (Fuller et al. 2022). Researchers estimate the annual number of victims of this pollution to be 9 million, or 16% of total deaths on the planet. The reduction in premature deaths from water poisoning or indoor air pollution is undeniable in recent decades, but this decrease has been largely offset by the increase in deaths linked to outdoor air pollution and exposure to toxic substances directly induced by contemporary industrialization and urbanization (these deaths have increased by 66% since 2000). More than 90% of the victims of environmental pollution live in developing countries, but examining the data relating to air pollution in Europe makes it possible both to understand the importance of this challenge for developed countries and to highlight the role of social inequalities in ecological vulnerability (which is the product of exposure to a nuisance and the sensitivity of exposed populations).

In spite of past efforts, air pollution remains a major health challenge for Europeans as well: in the most recent study to date by the WHO's Europe bureau, experts note that "air pollution is the largest environmental health risk in Europe", an assertion confirmed by the European Environment Agency, whose latest assessment similarly states that "air pollution is the biggest environmental health risk in Europe",

[7] This increasingly visible intersection between the health issue and the ecological challenge was recalled on October 11, 2021 by the WHO in its report published for COP 26 ("Acting for the climate in the name of health") and accompanied by a call for 45 million health practitioners to sign a "climate prescription".

Table 2.2 Health impacts of climate change

Climate change fueled by the consumption of fossil fuels					
Acidification of seas and oceans		*Increase in average temperatures and extremes (modification of the configuration of precipitation; sea level rise; weather extremes)*		*Pollution related to fossil fuels*	
Decrease in productivity of some fishing and aquaculture; proliferation of toxic seaweed	Collapse of biodiversity and ecosystems	Floods; drop in agricultural returns	Heat wave; drought; fires	Increase of ozone	particulate pollution; atmospheric pollution; pollen
Under- nutrition; bacterial diarrhea	Diseases transmission vector	Cardiovascular diseases	Respiratory diseases	Mental stress	Allergic pathologies

Source Adapted from IPCC (2022)

with "almost all Europeans still suffering from air pollution, leading to about 400,000 premature deaths across the continent".[8] The OECD similarly stresses the magnitude of the health challenge of air pollution in Europe: "depending on the methods of estimation, between 168,000 and 346,000 premature deaths across all EU member states in 2018 can be attributed to exposure to outdoor air pollution in the form of fine particles ($PM_{2.5}$) alone. This represented 4–7% of all deaths in 2018. In addition, hundreds of thousands of people develop various illnesses associated with air pollution, leading to a loss of about 3.9 million disability-adjusted life years (DALYs) annually in the European Union." In fact, exposure of Europeans to ambient air pollution appears to be a key element of the perception of the quality of life within the European population.

This health-environment impact has of course an intergenerational dimension: a recent study (Wim Thiery et al. 2021) shows that a person born in 1960 will experience an average of about 4 ± 2 heat waves in their lifetime; in contrast, a child born in 2020 will experience 30 ± 9 heat waves in a scenario determined by current climate commitments, which could be reduced to 22 ± 7 heat waves if warming is limited to 2 °C, or to 18 ± 8 heat waves if it is limited to 1.5 °C. In any case, it is four to seven times more than for a person born in 1960.

Environmental factors are in fact more important for children than for adults. According to WHO, in 2012, 1.7 million deaths of children under five were attributable to the environment (these included 570,000 deaths from respiratory infections, 361,000 deaths from diarrhea, 270,000 deaths from neonatal conditions, 200,000 malaria deaths and 200,000 accidental trauma deaths).

Already, life expectancy is being stalled by a number of social and environmental factors: in some high-income nations, the long trend of improving life expectancy and health stalled after around 2012, and in some countries as a whole or for some groups,[9] it has actually

[8] According to the EEA, exposure to fine particulate matter caused 379,000 premature deaths in EU-28, where 54,000 and 19,000 premature deaths were attributed to nitrogen dioxide (NO_2) and ground-level ozone (O_3), respectively (EEA 2018).

[9] The Marmot Review (Marmot et al. 2020) showed that for the first time in more than 100 years life expectancy has failed to increase across the UK, and for the poorest 10% of women it has actually declined.

fallen (Fenton et al. 2019).[10] France, a poster child for life expectancy excellence, is a case in point (Box 2.1).

Box 2.1: Life Expectancy in France
The French healthcare system is considered to be among the best in the world, as shown for example by the effective universal coverage indicator which ranks France among the top five nations in the world[11]: thanks to a strong improvement in this indicator between 1990 and 2010, the index of effective universal health coverage (UHC),[12] which aims to measure the coverage rate of services contributing to the health needs of a population, increased from 66 to 88% from 1990 to 2010 then from 88 to 91% from 2010 to 2019.

Yet life expectancy has slightly declined in the decade from 2014 to 2022 (this downward trend being even more marked for life expectancy at age 60). This downward dynamic over a period of almost ten years contrasts with all the previous developments over an equivalent time step (life expectancy at birth increased by 3.5 months per year on average during the period 1946–2014).

Two questions then arise: how to explain this decline in life expectancy? Can we anticipate that it will continue in the future?

On the first point, two years are particularly noteworthy in the almost decade that has passed since 2014: the year 2015 and the year 2020. In 2015, for the first time since 1970, a decline in life expectancy was measured in nineteen OECD countries, which is attributed to a particularly severe flu epidemic which notably claimed the lives of tens of thousands of elderly and frail people. The largest reductions in life expectancy were seen in Italy (0.6 years) and Germany (0.5 years), erasing the equivalent of two years of gain. France then recorded a drop in life expectancy of 0.3 for women and 0.2 for men.

[10] Fenton L, Minton J, Ramsay J, M K-B, Fischbacher C, Wyper G, et al. (2019), "Recent Adverse Mortality Trends in SCOTLAND: Comparison with Other High-Income Countries." *BMJ Open*.

[11] Lozano et al. (2020), "Measuring Universal Health Coverage Based on an Index of Effective Coverage of Health Services in 204 Countries and Territories, 1990–2019: A Systematic Analysis for the Global Burden of Disease Study 2019", *The Lancet*, Volume 396, Issue 10258, pp. 1250–1284, https://doi.org/10.1016/S0140-6736(20)30750-9.

[12] The UHC effective coverage index aims to represent service coverage across population health needs and how much these services could contribute to improved health.

In view of the years that have passed since then, if the year 2015 appears to be strategic, it is because it intertwines two phenomena that can be described as "natural": the entry into advanced age of the baby-boom generation; and the impact of a seasonal virus (this was also the case in the year 2003, which intermingled the deadliest natural disaster since 1900 and the peak of the effect of "slack classes" on the reduction of annual deaths). The combination of these two phenomena therefore connects a social structure and an ecological shock, or rather the effect of a violent ecological shock on a fragile social structure. The same combination is responsible for an even more pronounced drop in life expectancy in France in 2020 (from 2019): 0.5 years for women and 0.6 years for men. But contrary to common perception, life expectancy has not since resumed its inexorable ascent: rather, it has found a new diminished trajectory.

The year 2022 is remarkable in this respect: 667,000 people died in France, only 2000 less than in 2020. In fact, the breakdown of deaths for the year 2022 is particularly intriguing when compared to the last normal year available (2019): +29,000 due to aging and population growth, –21,000 due to downward trend in mortality probabilities and +46,000 difference between expected and observed deaths.

The first two phenomena work in the opposite direction and result in a net increase of 8,000 deaths. The structural increase in deaths in France (planned, understandable and explainable), initiated since 2005, remains no less impressive: in 2022, only 56,000 units or 8% difference between the number of births and the number deaths in France, an extremely small gap between the dynamics of life and death that has not been seen for 70 years.

There remain the 46,000 so-called "excess" deaths which bring down life expectancy, more deaths in 2022 than in 2021, a year marked more strongly by the COVID-19 pandemic. This figure testifies above all to the combination of the violence of the heatwaves of the summer of 2022 (which claimed nearly 11,000 lives) and the comet tail of the COVID-19 pandemic.

Contrary to popular belief, France is not, and will not be, a haven of peace in the global ecological chaos: while the globe has warmed up by 1.1 °C since the pre-industrial period, France is warmed by 1.7 °C and Paris by 2.3 °C. In its annual report published on December 3, 2019, the Commission of the medical journal *The Lancet* offers an independent assessment of the effects of climate change on human health and concludes that France's vulnerability to the effects of heat on health is the "one of the highest in the world". France's vulnerability to ecological shocks in international comparison appears even more clearly when we consider human losses alone. According to the average Climate Vulnerability Index from 1998 to 2017 (Eckstein et al. 2020), the country obtains a score that places it 15th among the most vulnerable countries (out of nearly 180 countries assessed), the human losses per inhabitant observed placing it in 8th place in the world, by far the most affected European country (over the period 2000–2019, it ranks 27th for the overall index and 4th for the number of total victims).

This climatic risk is coupled with (and sometimes combined with) the deterioration of environmental living conditions which also threaten the well-being of the French population, starting with their health. Air pollution is in France (as in Europe) the most important environmental risk for human health. Recent studies suggest that this risk is still greatly underestimated for two major reasons: the first is the inadequacy of the air pollution danger thresholds adopted by the European Union, which are much higher than those of the World Health Organization (by way of illustration, only 4% of the European population is supposed to be exposed to dangerous pollution of fine particles according to the thresholds of the European Union, when 75% is exposed according to WHO thresholds; the second reason is the recent discovery of neurological (and not just respiratory) damage caused by the finest particles ("nanoparticles"). Thus, air pollution could be responsible for 100,000 deaths in France, double the official estimate, i.e., the equivalent of three waves of Covid each year, or even 15% of total deaths.

There is therefore every reason to think that this is not a cyclical phenomenon: life expectancy in France has probably entered a phase of precariousness under the impact of ecological shocks understood in the broad sense (viral and climatic, etc., shocks).

Since environmental risk factors (air, water and soil pollution, exposure to toxic chemicals, climate change, ultraviolet radiation) contribute to hundreds of diseases and health conditions, the concept of "co-benefits" seems intuitive: improving the environment (or helping to stop its degradation) is a lever for improving human health (Chapter 4).

Interest in environmental health, which is an essential means to improve health and the environment in a synergetic and holistic way, goes back to the work of Hippocrates and his *Treatise on Airs, Waters, Places,* in which he writes about the influence of environmental factors. But a key dimension should be added to environmental health: social justice.

Health and equity are indeed inextricably linked, as the editor-in-Chief of *The Lancet* Richard Horton eloquently explained while arguing that COVID-19 was a "syndemic" rather than simply a pandemic:

> Two categories of disease are interacting within specific populations—infection with severe acute respiratory syndrome coronavirus 2 (SARS-CoV-2) and an array of non-communicable diseases (NCDs). These conditions are clustering within social groups according to patterns of inequality deeply embedded in our societies. The aggregation of these diseases on a background of social and economic disparity exacerbates the adverse effects of each separate disease. COVID-19 is not a pandemic. It is a syndemic. The syndemic nature of the threat we face means that a more nuanced approach is needed if we are to protect the health of our communities. (Horton 2020)

A large body of research, compiled by Richard Wilkinson and Michael Marmot, has confirmed the negative impact of social inequalities on physical and mental health at local and national level (via stress, violence, reduced access to health care, etc.). Inequality also acts as a powerful driver of many diseases perceived as natural or biological in the developing world but which in reality result from unfavorable living conditions (unequal access to water, healthy food, etc.). The late Paul Farmer (1999) showed how inequality constitutes a "modern scourge" in terms of health as formidable as infectious agents.

Social and economic conditions and their effects on people's lives determine their risk of illness, the steps taken to prevent them from becoming ill or to treat their illness when it occurs. Among these "social determinants of health" (the circumstances in which people are born, grow, live, work and age and the healthcare systems to which they have access), environmental conditions play an important and growing role. Hence the notion of "social health inequalities", understood as a systematic relation between health and social category (be it gender, nationality, family composition, income, education, etc.).

Structural inequality also plays a major role in the face of extreme events driven by human activity (Smith et al. 2022). The resilience or, on the contrary, the vulnerability of populations also depends on the dynamics of social inequalities. Vulnerability to so-called "natural" disasters actually depends on exposure and sensitivity to a given shock, on the one hand, and adaptive capacity and resilience, on the other. Considered in this framework, inequalities increase exposure and sensitivity, and weaken adaptive capacity and resilience: they act as a multiplier of the social damage caused by environmental shocks for both developed and developing countries (such as was demonstrated by the human impact of the COVID-19 pandemic on societies deeply marked by inequalities such as the United States).

The link between health and inequality is thus clear: social determinants of health are the circumstances in which people are born, grow up, live, work and age, and the systems put in place to deal with illness. These circumstances are in turn shaped by a wider set of forces: economics, social policies and politics. The relation between environmental conditions, individual welfare and social outcomes is thus mediated by health issues and more generally by the impact of environmental conditions and policies on the well-being of individuals. A recent study (Rockström et al. 2023) has put forward the idea of "safe and just Earth system boundaries", updating the framework of "planetary boundaries" (biophysics laws) using human frontiers (justice principles), a combination that leads to alternative definitions of a "just transition" (see infra).

The Inequality Crisis

Economics is a very old discipline, whose inventors in the West were Xenophon and Aristotle some 2500 years ago. Economics appears for the first time as such in 380 BCE in the title of a concise treatise by Xenophon, a disciple of Socrates, which defines it as the art of administering an agricultural domain (the 21st chapter of Xenophon's book details the tasks and skills of a good "economist", i.e., a good landowner). The association of the root words *oikos* (home) and *nomos* (law or rule of conduct) is taken up by Aristotle in his *Politics,* published a few decades later, but this definition of the economy is richer than that of Xenophon: it is the discipline of sufficiency in the service of the satisfaction of essential human needs. It is therefore the discipline that reconciles the needs of humans with the constraints of their environment.

Aristotle clearly distinguishes two spaces of human socialization: the home and the city. Within the household, there is a correspondence between reasoned economic needs and limited natural resources. But this space of the home is by nature unfair: the members of the family are not equal and no authority aims to put them on an equal footing. The economic sphere, like the natural environment, is a world of injustice. In other words, there is no reason for the satisfaction of essential needs, which proceeds from a principle of necessity, to result in a just situation. It is in the city that the principles of justice are supposed to reign and lead to the formation of a circle of equality (a circle from which women, slaves and foreigners were excluded in ancient Greece). It is in the city that what is necessary will be deemed sufficient or not.

While this original relation between economic activity, natural resources and principles of justice appears intuitive, inequality had largely disappeared from economic studies before the renewal of inequality economics in the early 2000s. In fact, from the late nineteenth century until the very end of the twentieth century, the justice-centered political economy gave way to a so-called "economic science" focused on efficiency and largely blind to injustice. A new stream of mostly empirical economics has since the early 2000s shed light on the contemporary crisis of inequality (Box 2.2).

Box 2.2: The Income and Wealth Inequality Crisis
Three fundamental findings have recently been established regarding the state and evolution of income and wealth inequality within countries and globally.

First, Thomas Piketty and his co-authors have documented the rise of intra-national inequality (or "within-country" inequality): since the early 1980s, income inequality and even more wealth inequality have widened in all regions of the world, at different rates and from different initial situations (Europe remains a relatively egalitarian continent, the most unequal regions being South America, sub-Saharan Africa, the Middle East, and South Asia and Southeast Asia). Inequalities in wealth are more marked than income inequalities: in France, in 2018, the 10% of households with the highest incomes held 24.8% of all disposable income, while the 10% of households with the most wealth concentrated 46.4% of total wealth (the share of inherited wealth in total wealth now represents 60% compared to 35% at the start of the 1970s).

The second major contemporary development has to do with global inequalities: Branko Milanovic showed that, while internal inequality (within countries) was increasing, international inequality (between countries) had on the contrary started to decrease from 2000, but at a very slow pace. The Gini index is a measure of the concentration of income within the population: the closer it gets to 1, the greater the economic inequality. We can choose to weight (or not) this indicator at the global level according to the demographic weight of the countries of the world.

The Gini index of unweighted international inequality is around 0.55, the level of a country like Zambia, which is one of the most unequal in the world. But the Gini index weighted by population, giving more importance to very populous countries like China and India which have developed strongly over the past two decades, is around 0.45, roughly the level of inequality of the United States.

From 1820 to 1980, inequalities between countries increased constantly in relation to internal inequalities (from 10 to 55% of global inequalities), but this trend has now been reversed: in 2020, inequalities are greater within countries than between countries. In short, countries have converged, but people have not.

The neoliberal era that began in the early 1980s, therefore, seems to have resulted in two dynamics: on the one hand, the integration of the global South (in particular India and China) into international trade and financial flows, with the consequence of reducing the economic gap between the emerging countries and the countries of the Organization for Economic Co-operation and Development (OECD), and the widening of the gap between the richest and the other social groups in the emerging countries; and on the other hand, the decline in taxation on high incomes and companies and the decline of social protection (in short, of the social state) in OECD countries, with the consequence of widening the gap between the richest and other social groups within the developed nations and the stagnation of the standard of living of the middle classes.

Empirically, it is undeniable that intra-national inequalities on the one hand (Box 2.2) and environmental degradations and the consumption of natural resources on the other hand have increased simultaneously over the past four decades.[13] But how do the crisis of inequality and the crisis of the biosphere connect? The social-ecological approach (Laurent 2011, 2020) considers the reciprocal relationship between social dynamics and environmental dynamics by focusing on the intertwined nature of the twin crises that characterize the beginning of the twenty-first century. In this regard, the social-ecological approach works in two directions: social

[13] Studies and data relating to climate change, the destruction of biodiversity, the degradation of ecosystems and the consumption of natural resources, which are increasingly precise and solid, are now largely consensual: the climate is destabilizing rapidly and dramatically (IPCC 2021), ecosystems and nature's contributions to human well-being have been significantly degraded, biodiversity has been enormously eroded (IPBES 2019) and the consumption of natural resources is at a record level (IRP 2017).

inequalities fuel ecological crises while ecological crises in turn aggravate social inequalities.

Consider the climate crisis. On the one hand, a handful of countries, on the order of 10% (and a handful of people and industries within these countries), responsible for 80% of human greenhouse gas (GHG) emissions, are causing climate change that is increasingly visibly destroying the well-being of a considerable part of humanity in the world, especially in developing countries. On the other hand, the vast majority of the people most affected by climate change (in Africa and Asia), by the billions, live in countries which represent almost nothing in terms of responsibility but which are very vulnerable to the disastrous consequences of the climate change (heat waves, hurricanes, floods, etc.) caused by the way of life of the wealthiest.

The first causal arrow, which goes from inequality to environmental degradation, is an "integrated social ecology", because it shows that the differences in income but also in power between the rich and the poor (the powerful and the vulnerable) and the interaction of these two groups (and their multiple variations on the social palette) lead to increased environmental degradation and an acceleration of ecological crises that affect everyone within a group. For example, the freedom given to a coal-fired power plant in New Delhi to pollute the air by breaking environmental rules following the corruption of a local official will affect all the inhabitants near this plant regardless of their social status; this is even more true of the global pollution generated by this plant in the form of CO_2 emissions caused by the combustion of coal, which affects all the inhabitants of the planet, regardless of their level of income.

The reciprocal arrow of causality, which goes from ecological crises to social injustice, is a "differential social ecology", because it shows that the social impact of ecological crises is not the same for different individuals and groups, based on their socio-economic status. The polluting company in New Delhi will first affect the health of the most vulnerable because it will probably be installed in a poor neighborhood in which the inhabitants will not be able to oppose its activity; similarly, the climate change induced by the CO_2 emissions from this coal-fired power plant will first hit the most vulnerable and deprived countries and populations, including the vast majority of the Indian population). In the words of the aforementioned Brundtland Report: "As a system approaches ecological limits, inequalities only increase" (WCED 1987).

A good starting point to understand the need to connect inequality to environmental crises in order to grasp the holistic nature of sustainability analysis is to consider how three closely related and influential frameworks used to understand our ecological crises (the so-called "great acceleration", "planetary boundaries" and "tipping points") are lacking a solid social dimension.

The concept of the "great acceleration" is a set of 24 global indicators measured from 1750 to the present day and forming a "planetary dashboard" which makes it possible to identify a temporal breaking point around 1950. From this date, the influence of human socio-economic trends (population, economic growth, energy consumption, urbanization, etc.) is tangible on all components of the Earth's system in the form of GHG levels, ocean acidification, deforestation or loss of biodiversity. Humans have become biophysical agents. Hence the idea of humanity overcoming "planetary limits" which are quantitative thresholds "within which humanity can continue to develop and prosper for generations to come" (Steffen et al. 2015). Crossing these boundaries as we have done since 1950 amplifies the risk of causing large-scale, abrupt and/or irreversible environmental and social changes, which the literature now refers to as "tipping points" (McKay et al. 2022).

The authors warn us unambiguously of this peril, stressing that four of the nine planetary boundaries have now been crossed as a result of human activity: climate change, loss of biosphere integrity, Earth system change and altered biogeochemical cycles (phosphorus and nitrogen). Recently, a tenth limit has been identified as already exceeded: that of synthetic chemical pollutants including plastic (Persson et al. 2022), while the freshwater cycle has also been trespassed (Wang-Erlandsson et al. 2022). Two of these planetary boundaries, climate change and the integrity of the biosphere, are "core boundaries", the significant modification of which would "lead the Earth system into a new state".

But it is not possible to rely solely on biophysical indicators to determine when tipping points of overconsumption are reached, because ecological crises triggered by the destruction of the biosphere are not socially homogeneous in their consequences. The so-called "planetary boundaries" are really human frontiers: humans are no more equal in terms of vulnerability than they are in terms of responsibility when it comes to environmental damage. This is why, ultimately, sustainability is a matter of justice, and inequality matters both as a driver and an outcome of unsustainability. A socially differentiated ecological analysis, both in terms of causes and consequences, is essential for understanding the current ecological crises and their possible mitigation. Planetary boundaries must be understood as social-ecological boundaries in order to understand what social causes generate them and what social consequences they induce in return on social systems. We need to socialize our ecological knowledge.

In the social-ecological approach, we can more precisely define the Anthropocene as a time when social systems govern natural systems and are affected in return by the dynamics they set in motion. In other words, the domination of human societies over the biosphere has not put an end to their dependence on the biosphere. This is what Vitousek and his colleagues meant when they wrote more than twenty years ago: "It is clear that we control a large part of the Earth and that our activities affect the rest. In a very real sense, the world is in our hands and how we take care of it will determine its future and our fate" (Vitousek et al. 1997). We need to identify socio-economic tipping points to respond to biophysical tipping points.

So how can social inequalities affect the environment? Inequality is a cause and a consequence of the destruction of our environment: social inequalities catalyze ecological crises which in turn accelerate social inequalities.

Consider first, at the micro-ecological level, the behavior of the rich and the poor towards the environment in isolation. On the side of the wealthy, the sociologist Thorstein Veblen showed in his *Theory of the Leisure Class* (Veblen 1899) that the desire of the middle class to imitate the lifestyles of the more advantaged classes can lead to a cultural epidemic of environmental degradation. Veblen thus speaks of "conspicuous consumption", through which the desirability of the most harmful practices for the environment spreads in all social groups (driving the most polluting cars, living in the largest houses, consuming the most luxurious goods, etc.). On the side of the poor, Indira Gandhi explained in her speech at the first international summit on the environment in Stockholm in 1972 that "poverty and want are the biggest polluters". She wanted to show that, in the developing world, poverty leads to environmental degradation due to social emergency, for example the dramatic degradation of forest cover in Haiti and Madagascar due to the destitution of the population but also to multinational predation. Since the wealth of the world's poor resides primarily in natural resources (they lack access to other forms of wealth), the depletion of these natural resources leads to a spiral of impoverishment. The eradication of poverty is therefore desirable not only socially but also environmentally, provided that it does not take the form of a consumerist catch-up but instead is part of a redefinition and redistribution of global wealth.

Looking at the macro-ecological level, we can observe the effects of the interaction between rich and poor and its environmental consequences. We can then bring to light the mainsprings of the "political economy of the environment", a discipline which was largely shaped by Boyce (2002). The novelty of Boyce's approach is to study both current environmental challenges (such as climate change or air pollution) and current economic problems (such as unequal development and neoliberal globalization) through the prism of power inequalities and their corollary, income. It identifies the winners and losers of environmentally degrading activities, as well as the dynamics between the two groups, revealing how the winners are able to continue their activities by imposing their consequences on the losers.

Boyce distinguishes five different dimensions of power (Laurent and Zwickl 2021): purchasing power, decision-making power, agenda power, value power and event power. Purchasing power is classically defined as the ability to purchase goods and services produced in the economy. Decision-making power refers to the ability to decide on more or less environmental regulations. Distribution of agenda power refers to the ability to determine which topics receive the attention of the media and politicians, while value power affects how people's values and beliefs are shaped and changed by the public debate. Finally, event power refers to the ability to influence events that form the framework within which people must then make decisions.

In line with Boyce, one can identify four macro-ecological transmission channels of interaction between rich and poor through which social inequalities translate into harmful environmental consequences.

Inequality Increases the Need for Environmentally Harmful and Socially Unnecessary Economic Growth

Given that no country in the world has actually succeeded in decoupling (in absolute or net terms) economic growth from its negative impact on the environment (Parrique et al. 2019)—starting with carbon emissions (current growth is 80% based on fossil fuels) and waste—more economic growth will translate into more environmental degradation, whether locally or globally. Social inequalities therefore indirectly destroy the biosphere via the demand for growth.

Similarly, a recent study by Joel Millward-Hopkins (2022) shows that inequality dramatically increases the energy requirements needed to ensure a decent standard of living for all. A world where a decent life is universally assured, but where material inequalities within countries remain close to current levels, could experience twice the energy consumption of an egalitarian world with inequalities based solely on needs. By comparison, moving from the lowest plausible future population trajectory to the highest only increases global energy consumption in 2050 by about 18%.

But will the drop in emissions resulting from the impoverishment of the richest not be offset by the increase resulting from the enrichment of the poorest?

Some studies show that, globally, eliminating extreme income poverty ($1.90 a day) would only marginally increase global emissions, while capping all Gini coefficients of all countries in the world at 0.3 (close to that of the Nordic countries) would increase global emissions by less than 1% (Scherer 2018).[14] Further, researchers have attempted to answer precisely the question of whether the reduction in income inequality really leads to a reduction in CO_2 emissions (Millward-Hopkins and Oswald 2021). Their empirical analysis unfolds in two stages: the study first determines from existing surveys the levels of income inequality deemed fair in 32 OECD countries, and the researchers then show how these levels of inequality lead to lower carbon emissions inequality than current levels. But they insist on an important point: the total carbon footprint of countries does not change because of these lower income inequalities. In contrast, the combination of reducing inequality to fair levels and reducing total consumption by recomposition places the carbon footprint reduction burden on higher-income households, while the consumption of those at the bottom can decline (by reducing the luxury expenditure of the wealthiest groups).

[14] One needs to distinguish between the aggregation and political-economy effects of income redistribution. Aggregation leaves the relationship between income and environmental degradation per dollar unchanged; assuming the marginal impact per dollar diminishes with rising impacts, progressive redistribution increases total impact. The political-economy effect arises from shifts in the relationship so that redistribution may lead to lower impacts per dollar across the whole income spectrum.

Inequalities Increase the Ecological Irresponsibility of the Richest, Within Each Country and Between Nations

The wealthiest individuals and the wealthiest countries present a paradox. In surveys, they state that they are more concerned about the environment than the poor, and they are, according to these same surveys, more likely to adopt the best environmental practices or to favor more ambitious environmental policies. However, at the same time, they pollute more than the poor in absolute terms due to their higher incomes and more expensive lifestyles. They are also better able to protect themselves from the negative impacts of their behavior as they get richer (Laurent 2020).

Inequalities, Which Affect the Health of Individuals and Groups, Diminish the Social-ecological Resilience of Communities and Societies and Weaken Their Collective Ability to Adapt to Accelerating Global Environmental Change

A recent study shows that inequalities could in fact play a key role in the global ecological collapse. The study examines the possibility of a civilizational collapse, relying on a rich literature and relying on a new model called HANDY (human and nature dynamics), whose particularity is to add a variable of social stratification to the already existing characteristics of models that attempt to simulate terrestrial dynamics. Humans, in the model, are divided into "elites" and "commoners" and their consumption of natural resources is differentiated according to their economic and political power (Motesharrei et al. 2014).

The key idea of the model is that ecological collapse can occur not only because of ecological depletion of resources through overconsumption, but also because of economic stratification. Motesharrei et al. (2014) show, however, that this collapse through inequality can be avoided by a reduction in current levels of social stratification, a more equitable distribution of the consumption of natural resources and a greater efficiency of this consumption (although technological progress alone does not, in the model, prevent the eventual collapse).

Inequalities Reduce the Political Acceptability of Environmental Concerns and the Ability to Offset the Potential Socially Regressive Effects of Environmental Policies

All countries and localities that have adopted carbon taxes over the past two decades have also adopted offset mechanisms for households and businesses that have helped overcome initial resistance from citizens and businesses (as in the Nordic countries or in Indonesia). In the Canadian province of British Columbia, the carbon tax was rejected by 43% of its inhabitants when it was introduced without social offsets in 2008, to now be widely supported due to the implementation of social offsets. The "yellow vests" crisis in France (2018–2019) shows, in the same way, the need to combine environmental policies and the reduction of social inequalities.

These close relations and complex dynamics between inequality and the environment explain why a sustainability science "aiming at explaining nature-society interactions" should be focused on inequality and on actually "promoting equity" (Clark and Harley 2020).

But it should also highlight the need to connect inequality and cooperation: social distance diminishes trust and impedes the functioning of cooperation institutions such as schools, cities and the rule of law. In the light of the results of empirical studies presented by Bjørnskov (2007), microsocial factors seem to play a marginal role in determining the level of generalized trust, which really depends on a single macrosocial factor: income inequality.

The Cooperation Crisis

Human prosperity is primarily explained by social cooperation, i.e. our ability to act together to resolve our problems and fulfill our desires. Humans are cooperative animals.

In fact, the idea of laws of evolution reduced to mechanisms of fierce competition between selfish individuals for the transmission of the best genes is far too reductive and has long since been discredited. Division of labor exists in many species, as anthills or hives show.[15] Research

[15] British evolutionary biologist William Donald Hamilton even proved decades ago that individuals of certain species help members of their first circle to reproduce, which ensures an indirect form of gene transmission to the next generation (Hamilton 1964).

has recently highlighted the ability of certain insects not only to sacrifice themselves but also to sacrifice their ability to reproduce so that other individuals can perpetuate themselves. This is the case for "workers" among bees, ants or termites (Hamilton 1964, 1970; Bourke 2011). Collaboration between individuals of the same species is indeed a necessity for survival and reproduction: some dolphins who know how to hunt alone nevertheless decide to join congeners to implement a sophisticated technique aimed at enclosing their prey in concentric circles to maximize the volume of their grip.

All species in the biosphere work together for their survival or reproduction, some even in an altruistic mode, but they do so on a small scale, within the narrow circle of family or clan ties, and are unable to share collective intelligence to progress and develop. Humans can of course collaborate (work together for a specific goal) but they do much more than that: they have an ability to cooperate, i.e., an aptitude for boundless collective intelligence with unlimited purpose.

Cooperation is a crystallization of social ties that involves pooling and sharing intelligence in order to satisfy our needs and realize our desires. Trust is an expectation of reliability placed in human behavior that extends the operational field of cooperation in time and space. Cooperation and trust together form the heart of the specificity of humans since their origins: our species is said to be "ultra-social", its prosperity due to its capacity for cooperation. And this cooperation could not broaden and deepen without trust.

Since their appearance on Planet Earth roughly 7 million years ago, women and men have always cooperated through multiple channels of socialization, including the practice of singing and dancing around fires. However, this cooperation has become more and more extensive (its sphere constantly widening to encompass, far beyond blood ties, inhabitants of other cities, regions and countries) and sophisticated (its depth and intensity increasing, as with the invention of the great modern universities in the nineteenth century or on-line relationships in the early 1990s).

Studies aimed at understanding the origins and tenets of human cooperation have progressed a lot in recent decades, albeit in variously fruitful directions, to bring to light two fundamental observations. First, humans are unique among living species in the magnitude and the depth of their cooperation. Second, this cooperation, whether it is motivated by selfishness or altruism, is made possible by formal institutions (like laws) and informal ones (like rules of courtesy). Humans are not born cooperators, they become cooperators as they are immersed at an early age in these institutions of cooperation. And they do not find their inspiration in the living world: there is a fundamental break between the human species and others in the capacity that we have, not only to reproduce cooperative behaviors we observe among our elders, but to build lasting and improvable institutions useful for the cooperation of every human with all others, beyond blood ties.

The crisis of cooperation is not a novel phenomenon: in the middle of the twentieth century, the analysis of human collective action entered a phase of deep fatalism, if not depression. In the 1950s and 1960s, pessimism about the ability of social systems to generate cooperative behaviors and sustain them over time spread. Selfish human nature was not incriminated, but rather the perversity of institutions, reputedly discouraging cooperation, and even encouraging secession by giving too much leeway to individual strategies, by flattering the inclination of humans to social sabotage and by favoring their withdrawal into their own interests. Three major models of non-cooperation emerged in the social sciences: Kenneth Arrow's "impossibility theorem", Mancur Olson's "logic of collective action" and Garrett Hardin's "tragedy of the commons".

Economist Kenneth Arrow (1951) appears to be the pioneer of this first wave of cooperation pessimism. With his theorem of the impossibility of social choice, Arrow aims at demonstrating that there is no other solution than dictatorship (the imposition on everyone of the preferences of a single person) to the problem of the logical aggregation of the preferences of several individuals, that is to say collective choice under democratic constraint. His reasoning is as follows: several possible choices are available to individuals who must jointly determine a common action, for example a vote in an election or the choice of a tax reform. Is there, asks Arrow, a procedure for reconciling individual choices compatible with freedom of choice and consistency of choice (i.e., the logical ordering of preferences)? He answers in the negative: if we intend to reach a common solution, it is impossible to respect the two criteria, that of freedom and that of coherence—hence this theorem is called "impossibility".

The consequence is clear: if humans want to associate themselves in a common action, they will either have to renounce the logic of their preferences, which will therefore not be respected, or rely on a superior authority which will simplify for them the problem of cooperation by imposing its own choice on them (this is the case when tax rules are enacted by an authoritarian power, as in China today). In short, Arrow demonstrates the inadequacy of institutions in the face of the complexity of human cooperation: even the best-conceived institutional engineering proves incapable of allowing the passage from the multiple to the common.

For Mancur Olson, another American economist, not only are the deviant behaviors that make human cooperation impossible not prevented by the institutions of cooperation, but they are encouraged by them. In *The Logic of Collective Action*, published in 1965 (Olson 1965), Olson starts from the idea of a supposed continuity between, on the one hand, selfish individuals rationally pursuing their interests and, on the other hand, human groups rationally pursuing the interests of their members. In fact, he asserts, this continuity is a figment of the imagination: individuals endowed with reason and motivated by their own interests will not work in favor of the interests of the groups to which they belong.

Here, it is the possibility of secession and privatization of gains that makes cooperation impossible. Free-rider behaviors, non-cooperative strategies *par excellence*, depend on the incentives perceived by individuals who intend to minimize their costs and maximize their benefits within groups, these depending on the size of the groups. In a small group, such as a professional association (doctors, lawyers, opticians, etc.), individuals cooperate and contribute to the common interest. On the other hand, in the broader context of national politics, the group itself will tend to maximize its profits (by pooling its losses and privatizing its gains) to the detriment of the community. It will then behave like a lobby – a term generally used in a pejorative sense in the US. We can cite the example of the lobby which intends to defend the use of pesticides recognized as probable carcinogens. It is the freedom to harm, and not just to choose, that is at the root of Olson's fatalism. Even if there is consensus within the large group about what the common good is and how to achieve it, non-cooperative strategies will prevail. For Olson, non-cooperative behaviors are inherent to institutions.

Biologist Garrett Hardin took social fatalism a step further by showing that non-cooperative behaviors could lead to collective ruin. In a resounding article in 1968 deploring the "tragedy of the commons" (Hardin 1968) at work on a planetary scale, he develops an apparently intuitive reasoning. If we rely on the logic of the "invisible hand" of Adam Smith, according to which, in a market economy, individuals should concern themselves only with their personal interests, without ever worrying about the fate of others, the sum of the egoisms of each one will lead not to the "wealth of the nations" promised by Smith, but to the ruin of the citizens. The allegory chosen by Hardin is that of the shepherds exhausting the pasture they share without owning it, for lack of effectively dividing up its use. The logic is close to that of Olson: if each breeder intends to privatize his gains (the sale of well-fed cows on the market) while pooling his losses (the consumption of grass by more and more cows), the pasture will quickly be overexploited, the "invisible hand" ransacking the common resource. Breeders, like their animals, will quickly wither away. Since "injustice is preferable to common ruin", it is wise, according to Hardin, to institute "reciprocal coercion by mutual acceptance", in other words to resort to a central authority which monopolizes the power to choose (as with Arrow) and which strongly resembles a dictatorial government. Herders will bend to the rules of grazing for their own good. By constraining individual freedoms, dictatorship relieves humans of the headache of cooperation.

But in parallel to these pessimistic views, studies emerged highlighting how cooperation was spontaneously greater than theories predicted and how well-designed institutions could channel trust, accelerate cooperation, including environmental cooperation, and avoid "social dilemmas". Elinor Ostrom initiated the revival of the faith in collective action by fostering the revolution of the commons (cf. infra). But she was not alone in this endeavor.

In the fall of 1979, Robert Axelrod (1984), then professor of political science at the University of Michigan, decided to organize a seminar which took the form of a match[16] between the best game theory specialists of the time.[17] The professionals invited all asked themselves the same question: which cooperation strategy would prove to be the best in light

[16] Each participant was invited to submit to the organizers a strategy in the form of a computer program. Each program was pitted against the others over five games consisting of two hundred innings. This confrontation must determine the best possible strategy, the one that will allow you to pocket the maximum gains. Gains are distributed as follows: a program that chooses to cooperate while its counterpart decides to defect wins 0 points, while its counterpart wins 4; if the two programs opt for cooperation, they each gain 3 points; if both defect, they each earn 1 point.

[17] Game theory, invented by the American mathematician John von Neumann in the 1920s and refined in particular by John Nash in the 1950s, aims to determine, in situations of uncertainty or incomplete information, the best possible choice given that of others,

of the gains pocketed? The answer fit in the four lines of computer code devised by Russian-born American mathematician Anatol Rapoport. His strategy, which won the tournament hands down, was called "tit for tat". At the first move, the Rapoport program cooperates, then it adapts strictly to the move played by the partner: if the partner cooperates, it cooperates; if the partner defaults, he defaults; and it reverts to cooperation if the partner reverts to it after defecting.

It is therefore a mirror strategy whose mainspring is reciprocity – it would moreover be better called "strategy of cooperation by consideration". The experiment demonstrated that no universal rule builds cooperation better than paying attention to the actions of the other, including their negative actions. Its conception owes nothing to chance, because Anatol Rapoport hoped, like other researchers of his generation, that the development of game theory could put an end to armed conflicts, or at least contribute to preventing the nuclear apocalypse in the geostrategic context of the Cold War.

The "tournament" experiment of Robert Axelrod thus showed that humans manage to cooperate by exceeding their immediate personal interests (which, in the system of earnings of the tournament of Axelrod, would order them never to cooperate) because they educate themselves mutually to cooperation. But for that, they need a relatively long time horizon that is embodied in robust institutions: this cooperation by consideration cannot be built in one-off games with no tomorrow. It also presupposes a process of mutual discovery that cannot be reduced to the utilitarian fulfillment of a limited task. Cooperation is a game of collective intelligence whose outcome must remain indeterminate. This definition brings us to the current crisis of cooperation, which can be understood by shedding light on a key distinction between cooperation and collaboration.

As Adam Smith shows with the metaphor of the pin factory (where each worker becomes better by specializing in his task to optimize his productivity and therefore maximize production), humans have understood how to collaborate (Smith 2000). Economics is often understood as the discipline of collaboration, which allows us to understand why people work together and how, to improve again this fruitful association of a part, through its goals, it concentrates human energy on individual utility

studying "strategic interactions" between players of formal games, the most famous of which is the prisoner's dilemma.

(rather than on violence or contemplation). Through the power of the market, economic forces aggregate individual utilities to achieve collective efficiency.

Because they have been able to divide their common work (specifically, in eighteen sequences) and specialize each in the task for which they are respectively the best, workers in the pin factory make many more pins than if they had to assume alone all the stages of product assembly. The technical division of labor leads to increased productivity, which fuels the "wealth of nations"; from a maximum of 20 pins per day, the technical division of labor brings production to 4800. Labor, divided, has multiplied output by 240.

But this division of labor which isolates workers can become counterproductive. It actually contravenes the vision developed by Adam Smith himself in *Theory of Moral Sentiments* published some 20 years before *The Wealth of Nations*. In this first work, just as fundamental as the one that will earn him his worldwide fame, Smith bases liberal prosperity on "sympathy", that is, empathy, the ability to put oneself in another's shoes. This sympathy engenders confidence and altruism, in other words the ability not only to see the world with the eyes of others, but also desire the well-being of others without necessarily hoping to derive any benefit from it.

The eminently effective collaboration of the pin factory enriches humans in appearance, but it impoverishes them at the same time: the economic gain of productivity turns into a loss of social innovation. Smithian collaboration is an efficient division of labor with a definite goal, object and duration. Now, as Smith himself writes in *The Wealth of Nations*, "in a civilized society, [man] needs at all times the assistance and support of a multitude of men, while all his life would barely suffice to win him the friendship of a few people". He cannot hope for this help if he remains bound to social utilitarianism, which sees others as means and not ends.

The French founder of modern sociology Émile Durkheim (2002) understood how collaboration could interfere with social cohesion (or "organic solidarity" as he labels it). For Durkheim, it is the social division of labor that is at the heart of contemporary societies, and not, as with Smith, its technical division. The division of labor, far from separating women and men, reinforces their complementarity by leading them to cooperate. Everyone acquires through their work the feeling of being useful to the whole. Collaboration turns into interdependence, and interdependence into cooperation.

Durkheimian cooperation thus sets in motion not only the collective ability to organize the existing world, but also to invent a new one through the indeterminate game of cooperation. While Smith inscribes as a motto on the pediment of human collaboration: "Give me what I need, and you will have from me what you need yourself", for Durkheim we cooperate because we want it, but our voluntary cooperation creates for us duties that we did not want.

To cooperate, says Émile Durkheim, is "to share a common task". The idea of "sharing" is certainly the key to cooperation, but Durkheim's definition is too narrow if one understands by "task" a chore that must be distributed. To cooperate is to work freely in concert, including—this is the fundamental point—for a purpose other than survival, reproduction or labor. Where humanity stands out in the animal kingdom is in its unique ability not only to collaborate but also to cooperate to build, share and transmit common knowledge to present and future generations.

In short, if we collaborate to do, we cooperate to know. One might think that collaboration and cooperation are simply synonymous but five essential differences set them apart beyond their etymology (working together vs creating together):

1. Collaboration is exercised by means of work alone, while cooperation calls on all human capacities.
2. Collaboration is for a fixed term, while cooperation has no finite horizon.
3. Collaboration is a purposeful association, while cooperation is a free process of mutual discovery
4. Collaboration is vertical while cooperation is horizontal (labor can be forced but collective creation is always voluntary).
5. Collaboration aims at producing while cooperation aims at innovation (including innovating to produce less or no more).

Cooperation, not collaboration, is the source of human prosperity. At the level of society, collective intelligence provides the same functions as individual intelligence: it allows the ordering and synthesis of the existing world (organization), while bringing out the unexpected and the new (creation). Collaboration can fulfil the first function, but not the second.

Thus, the opposite of cooperation is not so much competition as secession (the fact of not wanting to cooperate) and defection (the fact of no longer wanting to cooperate).

Nothing is more evocative of the power of human cooperation than a classroom in which students actively participate, through the mediation of their teacher, in discovering their capacities beyond the content of a program or the learning of a subject. When we teach, we never quite know what we are teaching, that is to say what the pupils or students choose to learn from ourselves and themselves, and it is precisely this part of free will that allows transmission. The great virtue of the Finnish education system—which is always cited as an example, and rightly so—is in being not so much oriented towards collaboration as towards cooperation, fostered by autonomous teachers, left free in the methods, supports and content of their teaching.

Collective intelligence adventures such as the Human Genome Project (DeLisi 1988) allow us to understand why the greatest human advances are inaccessible to other living beings. The deciphering of the human genome was completed in June 2000, made public on February 12, 2001 and finalized on April 25, 2003. In about fifteen years, after several unsuccessful attempts, the human species has deciphered its biological nature by relying on its culture of cooperation. These adventures of collective intelligence are made possible by the free, cumulative and often accidental exploration of the frontiers of knowledge (a study project on the survivors of an atomic attack led to the reshaping of biology and medicine); they can only be achieved through large-scale social cooperation (here, national and international); and the latter feeds on the creation of new institutions (in other words, formal mechanisms of cooperation) that make these advances finally fruitful and form the framework for future advances in knowledge (in this case, the model of "open science").

Why do we choose to cooperate rather than going it alone? Two major explanations clash: on the one hand, cooperation would be a calculated sacrifice, consented to obtain a benefit (we respect social norms in order to fit into a human community from which we think we will be able to derive benefits); on the other, on the contrary, cooperation would be an altruistic but also rational act, aimed at satisfying the interest either of another individual or a community.

We can always refine these motivations and show that they fail to account for the diversity and richness of human behavior (some individuals do not cooperate for either of these reasons), or that they are complements rather than substitutes (humans cooperate for a mixture of these two reasons). But it must be recognized that these two explanations are based on interest and efficiency. Human cooperation, as I understand it, is a quest for shared knowledge. Humans actually cooperate for a benefit, but this benefit is knowledge, and its usefulness and effectiveness are unknown to them at the time they choose to cooperate rather than secede or defect. Far from being a social machine aimed at utility and efficiency, cooperation takes the form of an unbounded quest for collective intelligence. Trust is at the heart of social cooperation because it transforms uncertainty into risk and accelerates reciprocity between individuals.[18]

It is precisely the contrast between cooperation and collaboration that constitutes the contemporary crisis of cooperation: while collective intelligence is supposed to be greatly accelerated by the digital transition, it actually appears that the reign of collaboration harbors a crisis of cooperation. We are simultaneously living the reign of collaboration and the crisis, perhaps even the decline, of cooperation. Our high-frequency societies are also low-intensity societies. The hyperactivity of the brief digital encounter often exhausts the capacity of individuals to build lasting relationships. Our collaborative societies are frenetic but devitalized societies, nervous but unstable, and finally conservative, because incapable of innovation and adaptation.

The contemporary crisis of cooperation has two faces: the "epidemic of loneliness", which, more than the much-maligned rise of individualism, isolates people and prevents them from being part of society; and the "war against time", induced by a hypertrophied digital transition and a

[18] More precisely, trust is an expectation of reliability in human behavior, which presupposes a relationship with another human being (a relationship that can be mediated by a collective norm possibly embodied in an institution, in which case trust is based on respect for this norm), in the context of an uncertain situation (which includes the possibility of seeing the trust granted betrayed, the person who takes this risk placing himself in a position of vulnerability), in a specific purpose and context (not everyone can be trusted, at any point and at any time), this expectation of reliability being the fruit of an individual will (granting one's trust is a personal choice, even if it is often influenced by a social context). Trust can take different forms (trust between people, trust in institutions), be built according to various methods (familiarity, habit, calculation, culture) and have varying degrees (one can have weak or strong trust, blindly or absolutely not trust).

neglected ecological transition, making uncertain the future of cooperation under the combined pressure of an acceleration of the present and a darkening of the future.

Norbert Elias (1991) has shown how illusory it is to separate individuals from society, to the extent that they respond to rules of social behavior that they themselves have helped to construct, enact and implement. In this respect, it is increasingly apparent that current institutions contribute not to an exacerbation of individualism, but to an increasing isolation of people which undermines the foundations of social cooperation.

It is in the United States that the analysis of this new phenomenon has started under the label "epidemic of loneliness".[19] Harvard sociologist Robert Putnam foresaw the atrophy of "social capital" in American society in his book *Bowling Alone: The Collapse and Revival of American Community*, published in 2000 and based on an earlier article (Putnam 1995). Putnam argues that social capital is declining in the United States, this decline weakening political participation and civic engagement to ultimately endanger the vitality of American democracy itself.[20]

Before coming to the observation itself, it is important to conceptually distinguish between social isolation and loneliness. These two states are characterized by a disconnection of the individual from his sociability networks, but it is necessary to distinguish between the objective reality of isolation, measurable by certain indicators of the density of social contact (with social networks such as family, neighbors or work colleagues) and the subjectivity of the feeling of loneliness, measured by declarative means.

[19] There are two possible ways to understand this expression, which is particularly astute. The first, which takes it literally, refers to a contagion of loneliness, which is an oxymoron: the spread of an epidemic precisely presupposes encounter, otherwise transmission is impossible. In reality, this transmission takes place through institutions, which gradually isolate individuals. The second way to understand this "epidemic of loneliness" consists in considering not its mode of transmission, but its consequences, which are considerable in health terms. The term "epidemic" is then justified, since loneliness can be likened to a form of pathology or pathological vector that degrades human health.

[20] Based on the observation of the progressive depopulation of bowling leagues in the United States (the practice of bowling dates back to the seventeenth century and experienced a veritable golden age in the post-war decades), Putnam postulates the gradual but continuous decline of the social capital in the country. This leads to a fall in democratic participation and ultimately to a degradation of the quality of public life.

In light of new data and research, the situation described by Putnam at the turn of the millennium can be understood as the prologue to the crisis of sociability that undermines the United States and many high-income and developing countries today. A 2010 study shows that 42 million Americans suffer from chronic loneliness, or nearly 15% of the population. The Bureau of Labor Statistics survey of the use of American time allows precise measurement of the time of "socialization" and "communication" outside work and meals. Americans now spend exactly 39 minutes a day (compared to 45 minutes ten years ago) on both, compared to the time spent in front of screens—more than three hours. What is more, 30% of Americans live alone. The health consequences of this social isolation are well documented: it greatly increases the risk of premature mortality and contributes to the development of many mental and physical pathologies. Recent work, presented at the congress of the American Psychological Association in the summer of 2017, shows that, according to 148 studies on the subject covering 300,000 participants, the risk of premature mortality increases by 50% due to social isolation (in the case of France, studies show that people living alone are more likely to die prematurely than those living with a partner, the risk of death being quite simply twice as high between the ages of 40 and 50 for single people).

Very recent data allow us to deepen our understanding of the phenomenon. The General Social Survey tells us that, in 2018, for the first time in the existence of these data, a majority of Americans aged 18–34 (51%) reported not having a "stable partner"; they were 33% in 2004 (this same figure being almost stable between 1986 and 2004—around a third). Simultaneously, happiness and life satisfaction among American teens, which increased steadily between 1991 and 2011, suddenly decreased after 2012 (Twenge 2019). In 2017, American teenagers aged 17–18 spent more than six hours a day of their free time on just three digital media: the internet, social networks and text messaging. Two activities which have a decisive influence on happiness declined in parallel: the time of rest and socializing (seeing friends, going to parties). Robust studies indicate a clear link between the use of social networks and feelings of social isolation (see for instance Primack et al. 2017). In addition, numerous indicators drawing the contours of a new psychological fragility (depression, suicidal thoughts, self-harm) have increased sharply among adolescents since 2010, particularly among girls and young women (a trend accelerated by the Covid years; CDC 2023).

In France, as well, social isolation also massively affects young people. A recent report shows that 18% of young people, i.e., more than two million French people aged 15–30, are socially vulnerable to varying degrees, ranging from fragility to complete isolation (Fondation de France 2021). According to this study, young people who are isolated or socially vulnerable, who say they are not satisfied with their housing and transport conditions, are more likely than the average to be in a situation of school failure, more of them than other young people declare that they have no diploma and are less well integrated (they declare themselves inactive twice as often as all young people aged 15–30). They also have a poorer state of health than other young people in their age group. And yet, these isolated young people are fully immersed in new information and communication technologies and in the digital transition, devoting even more time to screens than all 15–30-year-olds, who themselves devote ten hours more per week than the rest of the population to it. This may explain it, but it is in any case an illustration of the danger of confusing digital connection and social connection. In the case of isolated young French people, digital connection manifests as a symptom of social disconnection.

The digital disconnection also affects work as "collaborative management" spreads worldwide and with it, a double social isolation contributing to the "epidemic of loneliness".[21] New work constraints are increasingly isolating individuals from their family and friends, in other words, from their network of sociability, while loneliness is also growing in the workplace itself. Recent work shows how the digital, far from abolishing human labor, degrades it and, in some ways intensifies it. What it really abolishes is leisure, by providing employees with the same tools for leisure and for work. This confusion reached its peak with the generalization of telework under the effect of containment policies and social distancing. The empire of labor no longer stops at the transportation door or the front door. What is called "blurring" is in fact an imperialism of labor.

While digital tools are supposed to connect us in new ways, they end up de-socializing humans. The war against time, especially free time, which we have declared through the digital transition is also a war against day-to-day attention and social cohesion in the long run. Thus a growing

[21] Today, on average in OECD countries, only 17% of employees are union members, compared to 30% 30 years ago; only a third of workers are now covered by collective agreements, whereas almost half were 30 years ago.

contradiction between the digital and ecological transition. At the very moment when social and ecological challenges require the mobilization of a maximum and continuous social energy, digital acceleration diverts our collective energy. The crisis of cooperation fueled by digital disconnection and growing inequality is also a crisis of sustainability.

References

Arrow, K. (1951), *Social Choice and Individual Values.* Cowles Foundation for Research, Yale University Press.

Axelrod, R. (1984), *The Evolution of Cooperation.* New York: Basic Books.

Bjørnskov, C. (2007), "Determinants of Generalized Trust: A Cross-Country Comparison", *Public Choice*, vol. 130, pp. 1–21. https://doi.org/10.1007/s11127-006-9069-1.

Bourke, A. F. G. (2011). "Principles of Social Evolution", Oxford University Press, Oxford.

Boyce, J. (2002), *The Political Economy of the Environment.* Cheltenham: Edward Elgar.

CDC (2023), "Youth Risk Behavior Survey Data Summary & Trends Report: 2011–2021."

Clark, W., and Harley, A.G. (2020), "Sustainability Science: Toward a Synthesis", *Annual Review of Environment and Resources*, vol. 45, no. 1, pp. 331–386.

DeLisi, C. (1988), "The Human Genome Project", *American Scientist*, vol. 76, pp. 488–493.

Durkheim, É. (2002 [1893]), *De la division du travail social*, Jean-Marie Tremblay (ed.), Classiques des sciences sociales, Université du Québec à Chicoutimi. http://classiques.uqac.ca/.

Eckstein, D., Künzel, V., Schäfer, L., and Winges, M. (2020), "The Global Climate Risk Index 2020. Who Suffers Most from Extreme Weather Events? Weather-Related Loss Events in 2018 and 1999 to 2018". German Watch.

Elias, N. (1991), *La Société des individus.* Paris: Fayard.

European Environmental Agency, (2018). "Air Quality in Europe". 2018 Report, Publications Office, Luxembourg.

Farmer, P. (1999). "Infections and Inequalities. The Modern Plagues", University of California Press, Berkeley.

Fenton, L., Minton, J., Ramsay, J., Kaye-Bardgett, M., Fischbacher, C., Wyper, G.M.A., and McCartney, G. (2019), "Recent Adverse Mortality Trends in Scotland: Comparison with Other High-Income Countries", *BMJ Open.* 31, vol. 9, no. 10, p. e029936. https://doi.org/10.1136/bmjopen-2019-029936.

Fondation de France (2021), «Les solitudes en France—Un tissu social fragilisé par la pandémie».
Fuller R., et al. (2022), "Pollution and Health: A Progress Update", *The Lancet Planetary Health*, May 2022.
Hamilton, W. D. (1964). "The genetical evolution of social behavior, I & II ", *Journal of Theoretical Biology*, vol. 7, no. 1, pp. 1–52.
Hamilton, W. D. (1970). "Selfish and spiteful behaviour in an evolutionary model", *Nature*, vol. 228, pp. 1218–1220.
Haidong, W., et al. (2022), "Estimating Excess Mortality Due to the Covid-19 Pandemic: A Systematic Analysis of Covid-19-Related Mortality, 2020-21", *The Lancet*, vol. 399, no. 10334, pp. 1513–1536.
Hardin, G. (1968), "The Tragedy of the Commons", *Science*, vol. 162, no. 3859, pp. 1243–1248. http://www.jstor.org/stable/1724745.
Horton, R. (2020), Offline: "COVID-19 Is Not a Pandemic", *The Lancet*, vol. 396, no. 10255, p. 874.
IPBES (2019), "Global Assessment Report on Biodiversity and Ecosystem Services of the Intergovernmental Science-Policy Platform on Biodiversity and Ecosystem Services", IPBES.
IPCC (2021). "Climate Change 2021. The Physical Science Basis. Contribution of Working Group I to the Sixth Assessment Report of the Intergovernmental Panel on Climate Change".
IPCC (2022), "Climate Change 2022: Impacts, Adaptation and Vulnerability. Working Group II Contribution to the IPCC Sixth Assessment Report".
IRP (2017). "Assessing Global Resource Use. A Systems Approach to Resource Efficiency and Pollution Reduction". A Report of the International Resource Panel, United Nations Environment Programme, Nairobi.
Laurent, É. (2011), "Issues in Environmental Justice Within the European Union", *Ecological Economics*, vol. 70, no. 11, pp. 1846–1853.
Laurent, É. (2020), *The New Environmental Economics. Sustainability and Justice*. Cambridge/Medford: Polity Press.
Laurent, E., and Zwickl, K. (eds) (2021), *The Routledge Handbook of the Political Economy of the Environment*. London: Routledge.
Lawler, O.K., et al. (2021), "The Covid-19 Pandemic Is Intricately Linked to Biodiversity Loss and Ecosystem Health", *The Lancet Planetary Health*, vol. 5, no. 11, pp. e840–e850.
Lozano, et al. (2020), "Measuring Universal Health Coverage Based on an Index of Effective Coverage of Health Services in 204 Countries and Territories, 1990–2019: A Systematic Analysis for the Global Burden of Disease Study 2019", *The Lancet*, vol. 396, no. 10258, pp. 1250–1284. https://doi.org/10.1016/S0140-6736(20)30750-9.
Marmot, M., Allen, J., Boyce, T., Goldblatt, P., and Morrison, J. (2020), "Health Equity in England: The Marmot Review 10 Years On". Institute

of Health Equity. https://www.health.org.uk/publications/reports/the-marmot-review-10-years-on.
McKay, D.I.A., et al. (2022), "Exceeding 1.5 °C Global Warming Could Trigger Multiple Climate Tipping Points", *Science*, vol. 377, no. 6611, p. eabn7950.
Millward-Hopkins, J. (2022), "Inequality Can Double the Energy Required to Secure Universal Decent Living", *Nature Communications*, vol. 13, no. 1, p. 5028.
Millward-Hopkins, J., and Oswald, Y. (2021), "'Fair' Inequality, Consumption and Climate Mitigation", *Environmental Research Letters*, vol. 16, no. 3, p. 034007.
Motesharrei, S., Rivas, J., and Kalnay, E. (2014), "Human and Nature Dynamics (HANDY): Modeling Inequality and Use of Resources in the Collapse or Sustainability of Societies", *Ecological Economics*, vol. 101, pp. 90–102.
Olson, M. (1965), *The Logic of Collective Action: Public Goods and the Theory of Groups.* Harvard University Press.
Parrique, T., et al. (2019), "Decoupling Debunked. Evidence and Arguments Against Green Growth as a Sole Strategy for Sustainability", European Environment Bureau (EEB), Bruxelles.
Pekar, J.E., Magee, A., Parker, E., et al. (2022), "The Molecular Epidemiology of Multiple Zoonotic Origins of SARS-CoV-2", *Science*, vol. 377, no. 6609, pp. 960–966.
Persson, L., et al. (2022), "Outside the Safe Operating Space of the Planetary Boundary for Novel Entities Environ", *Environmental Science & Technology*, vol. 56, no. 3, pp. 1510–1521.
Prados de la Escosura, L. (2015), "World Human Development: 1870–2007", *Review of Income and Wealth,* vol. 61, no. 2, pp. 220–247. https://doi.org/10.1111/roiw.12104.
Primack, B., et al. (2017), "Social Media Use and Perceived Social Isolation Among Young Adults in the U.S.", *American Journal of Preventive Medicine*, vol. 53, no. 1, pp. 1–8.
Putnam, R.D. (1995), "Bowling Alone: America's Declining Social Capital", *The Journal of Democracy*, vol. 6, no. 1, pp. 65–78.
Redvers, N., (2021), "The Determinants of Planetary Health", *Lancet Planet Health*, vol. 5, pp. e111–e112.
Riley, J. C. (2005), "Estimates of Regional and Global Life Expectancy, 1800-2001", *Population and Development Review*, vol. 31, no. 3, pp. 537–543.
Rockström, J., Gupta, J., Qin, D., et al. (2023), "Safe and Just Earth System Boundaries", *Nature*. https://doi.org/10.1038/s41586-023-06083-8.
Scherer, L. (2018), "Trade-Offs Between Social and Environmental Sustainable Development Goals", *Environmental Science & Policy*, vol. 90, pp. 65–72.

Smith, A. (2000), *An Inquiry into the Nature and Causes of the Wealth of Nations*. Library of Economics and Liberty. https://www.econlib.org/library/Smith/smWN.html.

Smith, G.S., Anjum, E., Francis, C., Deanes, L., and Acey, C. (2022), "Climate Change, Environmental Disasters, and Health Inequities: The Underlying Role of Structural Inequalities", *Current Environmental Health Reports*, vol. 9, pp. 80–89.

Steffen, W., Richardson, K., et al. (2015), "Planetary Boundaries: Guiding Human Development on a Changing Planet", *Science*, vol. 347, no. 6223, p. 1259855.

Twenge, J.M. (2019). "The Sad State of Happiness in the United States and the Role of Digital Media", *World Happiness Report* 2019.

Veblen, T. (1899), *The Theory of the Leisure Class. An Economic Study of Institutions*. New York: Macmillan.

Vitousek, P.M., Mooney, H.A., Lubchenco, J., and Melillo, J.M. (1997), "Human Domination of Earth's Ecosystems", *Science*, New Series, vol. 277, no. 5325, pp. 494–499.

Wang-Erlandsson, L., Tobian, A., van der Ent, R.J. et al. (2022), "A Planetary Boundary for Green Water", *Nature Reviews Earth & Environment*, vol. 3, pp. 380–392.

Watts, N., et al. (2021), "The 2020 Report of the Lancet Countdown on Health and Climate Change: Responding to Converging Crises", *The Lancet*, vol. 397, no. 10269, pp. 129–170.

Wim Thiery, et al. (2021), "Intergenerational Inequities in Exposure to Climate Extremes: Young Generations Are Severely Threatened by Climate Change", *Science*, vol. 374, no. 6564, pp. 158–160. https://doi.org/10.1126/science.abi7339.

World Commission on Environment and Development (WCED) (1987), *Our Common Future*. Oxford and New York: Oxford University Press.

Worobey, M., Levy, J.I., Serrano, L.M., et al. (2022), "The Huanan Seafood Wholesale Market in Wuhan Was the Early Epicenter of the Covid-19 Pandemic", *Science*, vol. 377, no. 6609, pp. 951–959.

CHAPTER 3

Vision: Holistic Sustainability

Framework and Visualization: The Social-Ecological Loop

Having identified our intertwined sustainability crises, it is time to detail analytically and represent visually the connections between planetary health, cooperation and justice that together constitute a new vision from which new sustainability policies can emerge.

As much as narratives are essential in transitions, they must be embodied in visions and ultimately representations. Economist Kate Raworth (Raworth 2018), the inventor of the "Donut Economy", devotes very convincing pages at the beginning of her book to the importance of visual representations in the history of economic thought, emphasizing the "power of pictures". This power is of course as old as culture itself: since the beginnings of humanity, the most powerful stories have been told with images and this power remains acute in our eminently iconographic era (videos, emojis, etc.).

The general framework of the social-ecological well-being vision consists in articulating ecological interdependence with social cooperation: humanity depends entirely on the natural world of which it is itself made up (all living beings have the same common ancestor) and the

É. Laurent, *Toward Social-Ecological Well-Being*, Palgrave Studies in Environmental Sustainability, https://doi.org/10.1007/978-3-031-38989-4_3

natural world now depends on humanity because of its power of domination (so-called "wild" species must now be preserved by human ingenuity in order to survive).

Humans are indeed organisms within an ecosystem (an ecosystem being a community of animals and plants living in interaction with their physical environment), the biosphere being the ecosystem of ecosystems.[1] By colonizing the biosphere, humans have formed "anthromes" (or "anthropogenic biomes"; Ellis et al. 2010) on the surface of the planet such as cities, villages, cultivated lands, pastures, etc. But humans are still part of nature: they are, like other forms of life, intimately and socially connected to their environment by flows of energy and nutrients: in order to be able to breathe and therefore live, we extract oxygen from the atmosphere by returning CO_2 to it and we collaborate in a million complex ways with our environment and the creatures that inhabit it, who themselves collaborate with each other.

Yet, the vision that has prevailed in the discipline of economics for the last four decades since the advent of the neoliberal era is that of a disjunction between social systems and natural systems (which are intuitively represented in the form of circles), the former taking resources from the latter and discharging waste into them (see Fig. 3.1a).

Then came the idea of bringing these two spheres together, their intersection being the space of "sustainable development" (a vision which emerged at the end of the 1980s). A third representation emerged with the "planetary boundaries" model at the end of the 2000s: social systems are embedded in the biosphere and their uncontrolled expansion threatens the security of humanity as threshold after threshold is being crossed. With the representation of the "Donut Economy" (in the mid-2010s), a fourth vision emerged: that of a balance to be sought between social needs (social floor) and ecological constraints (planetary boundaries) (Fig. 3.1b–d).

The vision of this book is of a fifth type: the social-ecological loop. The intuition behind it is straightforward: to preserve humanity from self-destruction, the major challenge of our early century is the reinvention of

[1] It is made up of biomes (different types of ecosystems characterized by certain climates, types of fauna and flora, etc., such as tropical rainforests or arid deserts), which themselves contain smaller ecosystems (like rivers, lakes, dunes, etc.), in which creatures live, including human beings (which are themselves ecosystems, notably home to hundreds of thousands of bacteria).

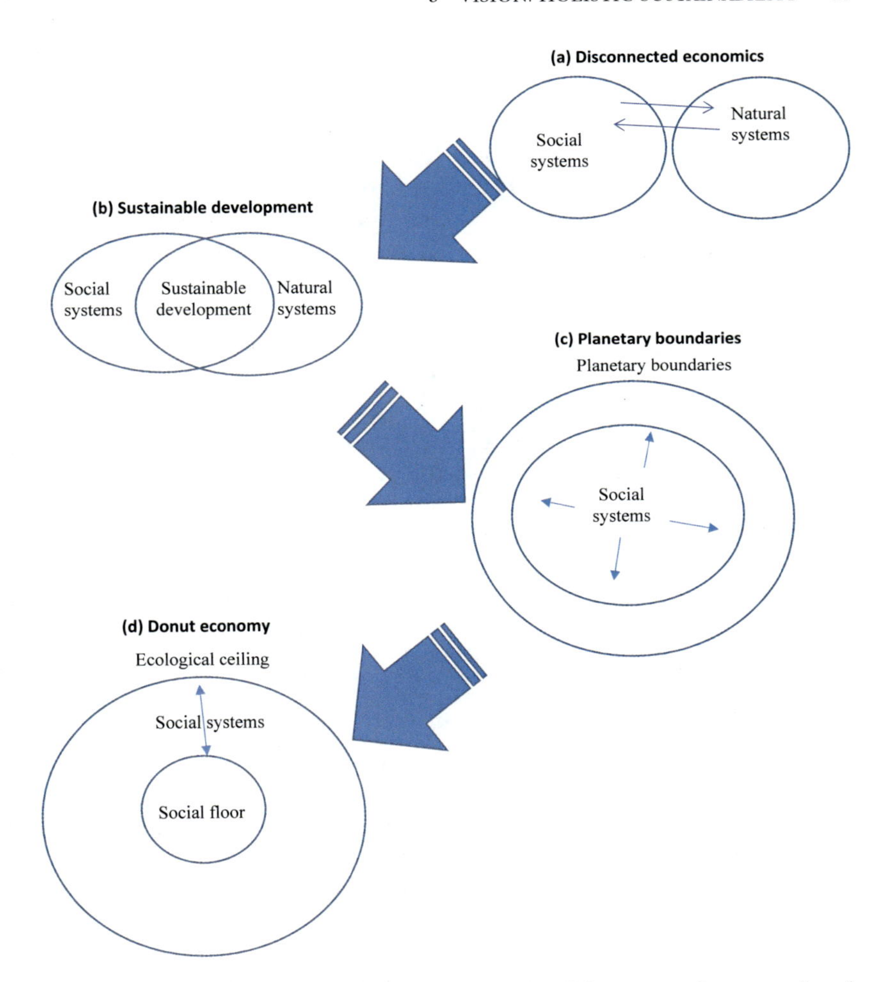

Fig. 3.1 **a–d**. Four social-ecological visions. **a**. Disconnected economics. **b**. Sustainable development. **c**. Planetary boundaries. **d**. Donut Economy

social cooperation in order to accomplish an ecological transition which consists mainly in preserving the living world on which human well-being depends entirely. Ecological interdependence allows social cooperation which in turn reinforces ecological interdependence: social and natural systems are thus intimately intertwined rather than simply juxtaposed or embedded. And this loop is dynamic: it can expand or shrink depending on the orientation of human development. The infinite intertwining of the two social and natural circles, which were separated at the origin of this iconographic journey (Fig. 3.1a), makes them whole.

The vision of the social-ecological loop articulates planetary health (ecological interdependence being the universal principle of the biosphere) and social cooperation (which is a specific human quality that depends on social justice) in a dynamic intertwining, which is endowed with three visual tools:

1. The key nodes of the systems (the link between inequalities and ecological crises; the link between ecosystem health and human health);
2. The width of the loop band that determines the direction of the whole social-ecological system: the larger the band, the more the system evolves toward sustainability;
3. The closeness of the two loops which describes how much humans live within or against the biosphere.

Let's start with the current situation, at the beginning of the twenty-first century, where both natural systems (planetary health) and social systems (cooperation) are visibly in a common crisis mediated by growing inequalities (Fig. 3.2).

The loop describes the intimate connections between how we treat the environment (the planetary health loop) and how we treat each other (the cooperation loop)—with inequalities of wealth and power as a blockage to the flow of well-being through both loops (inequality acting as a disease that constricts the flow of well-being, much as arteriosclerosis narrows the blood vessels).

Natural systems and social systems are connected by two powerful nexuses already explored in this book (Chapter 1): human health relies on the health of life within the biosphere but this unified health is degrading rapidly (environment-health nexus) and ecological transition policies are being hindered by the magnitude of domestic and international inequality that block or slows international cooperation and the adoption of environmental policies within countries and localities able to cope with current and future challenges (the inequality-unsustainability nexus).

Because of the non-transition (transition policies are not being adopted fast enough or at all), resource extraction and fossil-fuel energy

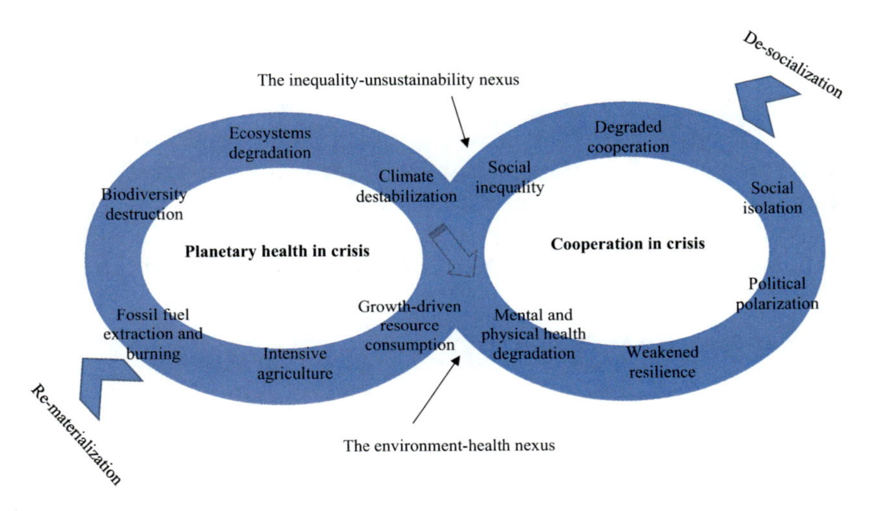

Fig. 3.2 Social-ecological crisis

consumption weaken the biosphere by increasingly destroying biodiversity, degrading ecosystems and destabilizing climate. The digital acceleration further increases the speed of resource consumption by re-materializing economic systems, which consume today three times the amount of natural resources they did 40 years ago, this consumption having accelerated with each digital innovation requiring more material and energy as they become ubiquitous.

These crises in natural systems are passed on to social systems via the degradation of physiological and mental health which have translated into weakened resilience to shocks, degraded cooperation taking the form of political polarization, social isolation, translating into even more transition hindering social inequality. Here also, the digital acceleration speeds up these trends by increasing social isolation rather than fostering social connection.

From the current situation, (at least) two futures are imaginable: one where those trends continue to the point where they lead to the attrition of both social and natural systems, effectively shrinking the social-ecological loop both in terms of the flow and contact between human and natural systems (Fig. 3.3a); the other where the vicious circle of the social-ecological loop is turned into a virtuous one, leading to an expansion of the flow within the loop and a wider contact surface between social and natural systems so that they tend to converge and even merge (Fig. 3.3b).

In this third, preferred scenario, planetary health, cooperation and justice are connected by three active principles: the first is that our world will be more just if it is more sustainable and more sustainable if it becomes more just (the just transition nexus); the second is that we can enjoy our social bonds as long as we take care of our natural links (the full health nexus); the third is that social cooperation feeds social innovation

(a) Social-ecological attrition

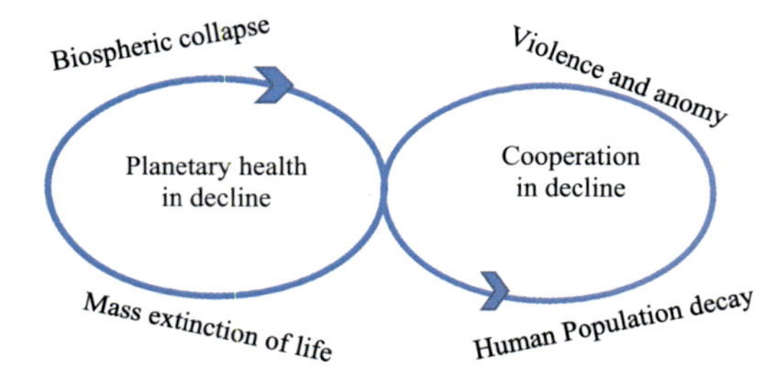

(b) Social-ecological revival

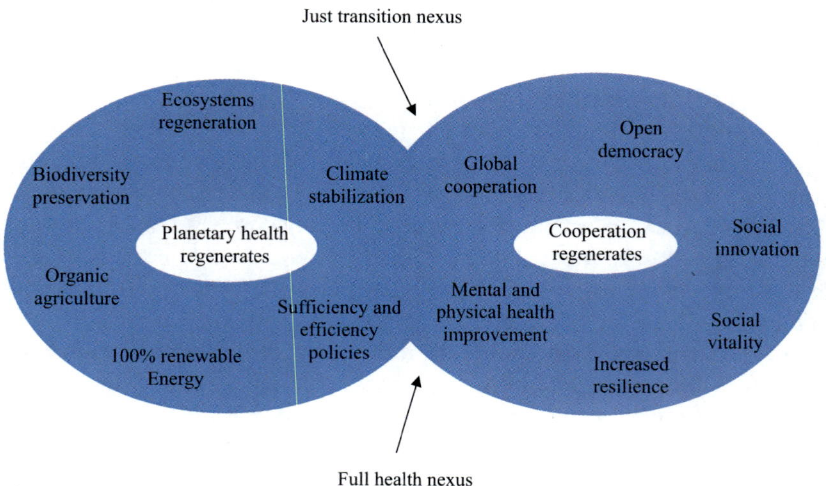

Fig. 3.3 **a**. Social-ecological attrition. **b**. Social-ecological revival

at the service of ecological interdependence (social-ecological regeneration). In this scenario, digital deceleration (i.e., slowing the pace of digital tools consumption and network use) fosters social-ecological revival. Let's explore in detail what those three dynamics entail.

Social-Ecological Well-Being in Motion: Full Health, Just Transition, Social-Ecological Regeneration

Full Health

"Are we able to reimagine a world where economies are focused on health and well-being?" With these words, the WHO issued a call to governments and citizens around the world on World Health Day, 7 April 2022, which marked the 74th anniversary of its founding. But this call does not mention the need to update the very definition of health with which WHO operates.

The preamble of the 1948 WHO Constitution defines health as "a state of complete physical, mental and social well-being and not merely the absence of disease or infirmity". In the face of the environmental crisis threatening health explored in this book, it seems necessary to update this definition, and define "full health" as a *continuous state of well-being: physical and psychological, individual and social, human and ecological.*

Full health first underlines health solidarity between humans: my health cannot flourish by degrading yours (this social dimension of health is especially salient in times of epidemics and pandemics). But the notion of "full health" goes even further to include the health of the ecosystem that sustains our own. Full health is therefore understood as the health of a humanity fully aware of the vital importance of its social links and environmental roots. Full health refers to human health in all its dimensions and ramifications (physiological, psychological, social, ecological). The important thing about this definition is to emphasize the holistic nature of the approach, the continuity of health, which links mental health to physiological health, individual health to collective health and human health to planetary health. Full health is therefore health based on interfaces, synergies and solidarities. In fact, the English notion of "health" shares the same Indo-European root as the French word *holistique*, which refers to the totality of a phenomenon or an issue.

The COVID-19 pandemic has shown how health is a collective matter that is blurred and distorted by calls for "individual responsibility". But the collectivity that we must take note of and become partners in goes far beyond the human race alone. All the countries where so-called individual responsibility has formed the basis of the response to Covid (Sweden, United Kingdom, United States, etc.) have had to back down precipitously in the face of the extent of the disaster. Countries considerably less well endowed with hospital resources have been able to successfully implement community health methods (Thailand, Senegal or Bhutan).

This integrated approach to human health in the twenty-first century has been explicitly recommended by the "One Health" panel of high-level experts convened in November 2020 under the aegis of the World Organization for Animal Health (OIE), the Food and Agriculture Organization of the United Nations (FAO), the United Nations Environment Program (UNEP) and the World Health Organization (WHO). Human health, animal health, plant health and environmental health, these experts tell us, are complementary and interdependent. The panel recently offered a definition of "one health"[2]:

> *One Health is an integrated, unifying approach that aims to sustainably balance and optimize the health of people, animals and ecosystems. It recognizes the health of humans, domestic and wild animals, plants, and the wider environment (including ecosystems) are closely linked and inter-dependent. The approach mobilizes multiple sectors, disciplines and communities at varying levels of society to work together to foster well-being and tackle threats to health and ecosystems, while addressing the collective need for clean water,*

[2] This approach was adopted as the key concept in the 2022 COP 15 Framework (Kunming-Montreal Global biodiversity framework), the framework acknowledging the interlinkages between biodiversity and health and the three objectives of the Convention. The framework is to be implemented with consideration of the One Health approach, among other holistic approaches that are based on science, mobilize multiple sectors, disciplines and communities to work together, and aim to sustainably balance and optimize the health of people, animals, plants and ecosystems, recognizing the need for equitable access to tools and technologies including medicines, vaccines and other health products related to biodiversity, while highlighting the urgent need to reduce pressures on biodiversity and decrease environmental degradation to reduce risks to health, and, as appropriate, develop practical access and benefit-sharing arrangements.

energy and air, safe and nutritious food, taking action on climate change, and contributing to sustainable development.[3]

Ecosystems are indeed essential not only to physical health, but also to mental health, as key determinants of happiness and social relationships and our capacities more generally (in the sense of Amartya Sen): our health (capacity to access adequate food, to avoid preventable diseases, to drink clean water, to breathe healthy air, access to energy sources protecting us from heat and cold), our security (capacity to live in a non-hazardous environment, to reduce vulnerability to ecological shocks and stress) and in general to be able to benefit from the essential elements for a good life (ability to access resources supporting our well-being). Hundreds of careful and robust studies[4] have mapped the benefits of biodiversity and ecosystems for health.[5]

Certain aspects of these interrelationships are not sufficiently well known, in particular the link between ecosystems and happiness. The authors of the World Happiness Report 2020 identify three essential links between ecosystems and mental satisfaction (Helliwell et al. 2020): first, "biophilia", which postulates an instinctive and close bond between human beings and other living organisms or habitats resulting from biological evolution. We love ecosystems and beings of nature in a selfless, non-instrumental way, like more or less close relatives. Then, natural environments can have indirect positive effects on our happiness, by

[3] The Food and Agriculture Organization of the United Nations (FAO), the World Organisation for Animal Health (OIE), the United Nations Environment Programme (UNEP) and the World Health Organization (WHO) validated this new definition on December 1, 2021 https://www.who.int/news/item/01-12-2021-tripartite-and-unep-support-ohhlep-s-definition-of-one-health. A more comprehensive approach can be found in the academic literature, including the five key underlying principles (see One Health High-Level Expert Panel [OHHLEP] 2022).

[4] These studies on health benefits from biodiversity and ecosystems are synthetized in Sandifer et al. (2015) and Frumkin et al. (2017).

[5] These include lower levels of negative emotions such as anger, frustration and sadness, reduced mental fatigue, stress and cortisol levels, reduced incidence of respiratory diseases such as asthma, reduced mortality from stroke, and increased physical activity, happiness and self-esteem, as well as many other cognitive, psychological and physiological benefits. This is in addition to a vast category of other benefits: social (e.g., easier interaction), economic (e.g., increased value of properties surrounding areas such as parks) and spiritual (e.g., increased inspiration).

encouraging certain behaviors, e.g., exercise physical or social interactions, which improves the mental or physical health and longevity, and by consequence subjective satisfaction. Finally, natural environments can be free from certain human stressors, such as atmospheric or noise pollution, which are associated with respiratory and cardiovascular diseases, while they offer us "environmental amenities" (the environmental amenities can be defined as benefits or amenities drawn from natural resources, such as beauty or air quality). On both scores, human satisfaction is improved.

But the notion of "one health" does not sufficiently convey the fact that health is positioned as an interface of human societies and ecosystems and should be the core of the well-being economy (WHO, 2020). This is why "full health" seems a preferable alternative.

As already noted, full health first means social health. When it comes to the co-benefits of investing in social relations, it should be noted that the link between the quality and density of social life and physical and physiological health is of remarkable robustness. The link between social connection and reduced appetite for material goods is less clear-cut but nevertheless also well established. There is therefore good reason to believe that Europeans living in societies centered on the quality of social ties would be in much better health (therefore more resilient to ecological shocks such as COVID-19) and less absorbed by material consumption (hence reducing their ecological footprint).

Addressing social isolation, for instance in a country like France, is an important part of such a policy strategy: it has been estimated that, in 2020, 7 million people or 14% of French people aged 15 and over were in a situation of social isolation (since 2010, a gradual and continuous increase in the phenomenon has been observed, social isolation having increased by 5 points in ten years (from 4 to 7 million people).[6] Understood not as a choice of lifestyle, but as an insufficient connection to social networks, or even a total disconnection from sociability, social isolation is

[6] Credoc (2020).

growing in strength in a number of developed countries (such as the UK,[7] the US and Canada) with strong health-environment consequences.

It is, for instance, a risk factor in case of heatwaves (Klinenberg 2002 has shown how social relations effectively protect individuals during heatwaves). Investing in social relations and curbing social isolation (which can translate into policies such as increasing family and social time, investing in accessible childcare, fostering intergenerational relations, etc.) thus makes sense from the point of view both of mitigation of and adaptation to environmental shocks.

This holistic approach resonates and converges with contemporary reflections on the importance of developing preventive medicine rather than a purely curative medicine: mitigating social isolation through social-ecological prevention policy generates gains in well-being, measured by the alleviation of the burden of morbidity,[8] which leads to social revitalization being considered as part of public health.[9]

But full health is also a new horizon for social and even economic policy, a horizon which differs from full employment, set out in the 1944 Beveridge Report,[10] "Full Employment in a free society" as the post-war roadmap for European countries and the United States. William Beveridge, an economist by training, defines full employment as a situation "where the number of vacant places [is] greater than the number of job applicants, and [where] the places [are] such and localized in such a

[7] In the UK, the rise of social isolation and the loneliness that accompanies it have become a major political concern, to the point of having recently justified the creation of a "Loneliness Minister" straight out of a George Orwell novel. The social reality is indeed alarming: a 2017 report (stemming from the Jo Cox Commission on Loneliness) showed that 9 million people often or always feel lonely (around 15% of the population). The weakening of traditional places of sociability (unions, religious congregations, local shops and in particular pubs and companies) leads to the emergence of a "disconnected society", or, better, of a society isolation ("It seems our best friend is our smartphone," noted Rachel Reeves, who coordinated the report).

[8] Pantell et al. (2013).

[9] Snyder-Mackler et al. (2020).

[10] A report sometimes referred to as the second Beveridge report. The first report, released in 1942, is a detailed proposal for in-depth reform of the British welfare system towards which Beveridge has been thinking and working since its beginnings in the Webb Commission of 1905–1909 (named after its progressive president Beatrice Webb). The central proposition of the commission summarizes well the philosophy, revolutionary for its time, of the first social laws: "to guarantee a national minimum of civilized life".

way that unemployment is reduced to short waiting intervals" (referred to as "frictional unemployment").

The central objective of the Beveridge Report was clearly stated and quantified: reduce the unemployment rate to 3%, a task that only the state could perform, according to the text. This objective itself was explicitly conditioned to a balance of power favorable to employees, linked to an increase in productivity that depends on progress in manpower training. It is therefore a real system of social progress that Beveridge envisions.

The 1944 report is also known as the second Beveridge report. Finnish researcher Tuuli Hirvilammi (Hirvilammi 2020) recently recalled in this regard that the full employment system had been conceptualized as a "virtuous circle" by Swedish economist Gunnar Myrdal, thinker and craftsman of social protection in his country and theoretician of the welfare state. Myrdal's "virtuous circle" aimed to formalize the alliance between social protection and economic growth. This circle is virtuous because of two feedback nodes: full employment and education and training policies that link the wage level and labor productivity. The social-economic covenant, typical of the second half of the twentieth century in Europe, is cumulative: economic growth fueled by rising productivity increases employment which in turn nurtures social progress through the reduction of inequalities and the extension of social protection in all areas of the life cycle (education, housing, employment, pensions). Attitudes and behaviors (political trust, aspirations for social progress, etc.) propagate the structural dynamic. In this balance lies a good part of what we call social democracy.

The full employment system was itself part of a broader geopolitical framework, to which the expression "free society" refers directly: it asserted the superiority of the democratic camp over the communist sphere. While the Beveridge Report surprisingly recommends contemporary economic planning (the "socialization of demand" transferred from war to peacetime), it remains fundamentally committed to free enterprise and private ownership of the means of production, but also to the mobility of workers. In other words, the Beveridge Report is the manual for what will later be called in Germany the social market economy.

This is the first major difference between the system of full employment in 1942 and that of full health I propose here 80 years later: the geopolitical framework polarized from 1944 became a unified biospheric framework. The central issue today is no longer the ideological rivalry between two blocs competing for global supremacy, but the global ecological uncertainty that affects in one way or another all the nations of the planet (geopolitical tensions of course have not disappeared from our world).

The second and most substantial difference is that the objective of the social system is no longer full employment but full health. It certainly does not mean that the social utility of employment has disappeared, nor that we now live in societies freed from work, as we once were able to imagine. It is rather that our common ambition should no longer be to produce but to endure.

Full employment, under the previous system of social progress, now serves as a screen in the United States for a system, unfortunately extremely effective, of social regression. This conception of full employment—in reality unceasing work or "full labor"—clearly opposes to the health of individuals: to earn a decent income in the United States today, workers without qualification occupy three or four jobs with no benefits and numerous penalties (diseases, chronic pain, reduced life expectancy).

Conversely, health, and more specifically hope of life, does not disregard working conditions. Beveridge himself listed full employment within a system of social protection in which the improvement of health occupied a large place, disease being considered alongside poverty, insalubrity, ignorance and unemployment as one of the five "giant scourges" hampering progress of the working classes. Invoking the productivity gains that can be generated by a population of healthy workers, in terms of both physical and mental health, helped to convince his detractors, conservative politicians of the time, who would end up by being won over with the creation of the National Health Service in 1948. But health is for Beveridge an instrument of employment and economic growth, whereas today it must become an end in itself, beyond the economic growth that now stands in the way of social-ecological progress.

This old social-economic alliance, emblematic of the post-war period, has been destabilized by the rise of neoliberalism since the end of the 1970s. But that is not the most important change: the more or less robust links between economic growth and social protection were mainly played out in closed circuit, without consideration for the biosphere, even as human systems gradually became unsustainable in the last quarter of the twentieth century. This is why there is an urgent need to evolve from the old alliance between the economy and the social spheres which dangerously ignored ecology towards a new alliance between the social and the ecological which puts the economy in its proper place.

On the human side, full health combines the biological, mental, physiological and psychological dimensions and brings into play all the major questions of well-being: subjective happiness determined above all through social ties and natural ties, the social inequalities that weigh on the health of individuals, the solidity of the infrastructure that supports ecological aspects of human functions—breathing, drinking, eating—which naturally leads us to the supporting walls of our societies and our economies, the ecosystems.

On the side of natural systems, biodiversity (its abundance and its resilience) underpins the proper functioning of ecosystems and the benefits humans derive from them. This logic prevails of course in the opposite direction: when humans are destroying biodiversity, as they are doing massively today through their agricultural systems, they degrade these benefits and, at the end of the chain, damage their living conditions. The case of mangroves is one of the most telling: these maritime ecosystems promote animal reproduction, store carbon and constitute powerful natural barriers against tidal waves.[11] By destroying them, human communities become impoverished and weakened.

Full health should therefore be understood as the health of a humanity whose economic systems are embedded in the biosphere that gave them life, nourishes them and will carry them in its fall if it were to happen. Full health also reconciles the horizons of long and short time, because it is also the best resistance here and now in the face of ecological crises. It allows us to prepare for the future while facing the here and now.

Full health can rely on a number of existing concepts, metrics and data. At the second European conference on the environment and health held in June 1994 in Helsinki, a "Charter on the environment and health" was adopted. Its first article states: "Good health and well-being require a healthy and harmonious environment in which physical, psychological, social and aesthetic factors are given due importance. The environment should be seen as a resource for improving living conditions and increasing well-being" (World Health Organization Regional Office for Europe 1994). When updating this Charter in 1999, WHO experts considered that improving environmental conditions was the "key to better health".

[11] On this point, see Kathiresan and Rajendran (2005).

The WHO accompanied this conceptual recognition with a methodological innovation, by designing and popularizing an empirical method aimed at isolating the environmental factor in the "burden of disease", itself a new way of measuring the health of a population (Box 3.1).

Box 3.1: How to Measure Health
For quite some time, mortality indicators were exclusively used to assess the state of health of the population (infant mortality, premature mortality, life expectancy using mortality tables), but since the 1990s it is the question of morbidity and, more precisely, the "burden of disease" that has become central (Murray and Lopez 1996). This synthetic health indicator makes it possible, in the spirit of the WHO Constitution, to shed light on human health not as the inverse of death but as a state of well-being.

This transition from mortality to the burden of disease is reflected by the transition, in the 1990s, of the mortality rate to the DALY, an indicator which measures the years of life adjusted for disability by adding the years of life lost due to premature mortality and years of productive life lost due to disability caused by disease. (The DALY is equal to the sum of the number of deaths × standard life expectancy at age of death in years and the number of cases × severity of disability × average duration of the case until remission or death.)

It is thus possible to compare among themselves and for all groups all damage to health as a state of well-being and their respective impact.

The Global Burden of Disease survey is regularly updated and published in *The Lancet* and on the Health Metrics and Evaluation (IHME) website. The French health authorities have just launched a version of the burden of disease (Cnam 2021) using the most relevant national data to estimate the number of years of life lost (YLL), the number of years lived with a disability and the number of years of life corrected for disability.

For the year 2016, France had 586,519 deaths representing a total of 7.291 million YLL compared to the life expectancies estimated for the French population for the same year. The advantage of these indicators is that they make it possible to cross-reference the causes of morbidity with the demographic categories in order to precisely identify the pathologies of the groups most affected: in this case, we identify for France, on the one hand, the external causes of morbidity in young people (accidents and suicides, mental illnesses) and, on the other hand, tumors in older people, especially in men.

The WHO now estimates the environmental share of the global burden of disease at 24% and the share of total deaths due to environmental factors at 23%. According to the latest estimates published in March 2016, nearly one in four deaths worldwide resulted from having lived or worked in an unhealthy environment (children and the elderly being the most exposed to environmental risks). Here again, there is a strong inequality between geographical areas and population categories: the burden attributable to the environment is almost three times higher in Southeast Asia than in the American and European countries of the OECD (see Table 2.1).

It is actually possible to build an operational dashboard of full health, which can give rise to precise measurement, social appropriation by citizens and integration into public policies (Fig. 3.4 and Table 3.1).

How to design effective full health policies? To go back to the French healthcare system, its excellent performance masks the fact that it is too oriented towards the treatment of pathologies and the avoidance of mortality by therapy and insufficiently towards the prevention of the burden of morbidity, which, in a context of health fragility, can become a source of financial unsustainability and in a context of ecological shocks can become a cause of health vulnerability.

In other words, rather than constituting a sign of the quality of the healthcare system, the very significant production and consumption of healthcare and medical goods in France is partly fueled by a deterioration in French healthcare. Therapeutic excellence is necessary but not sufficient: it is also necessary to invest in prevention, particularly in view of the low preventive costs compared to increasing curative costs.

The weakness of prevention in the French healthcare system is notable in comparison with European neighbors, healthcare spending being largely devoted to inpatient and outpatient care rather than prevention. This is shown in Fig. 3.5: not only is the share of prevention in French health expenditure derisory, but it has fallen over the last decade, unlike in some of other countries. Finland in comparison has one of the highest shares devoted to prevention in the EU, more than twice the share of France (Figs. 3.5 and 3.6).

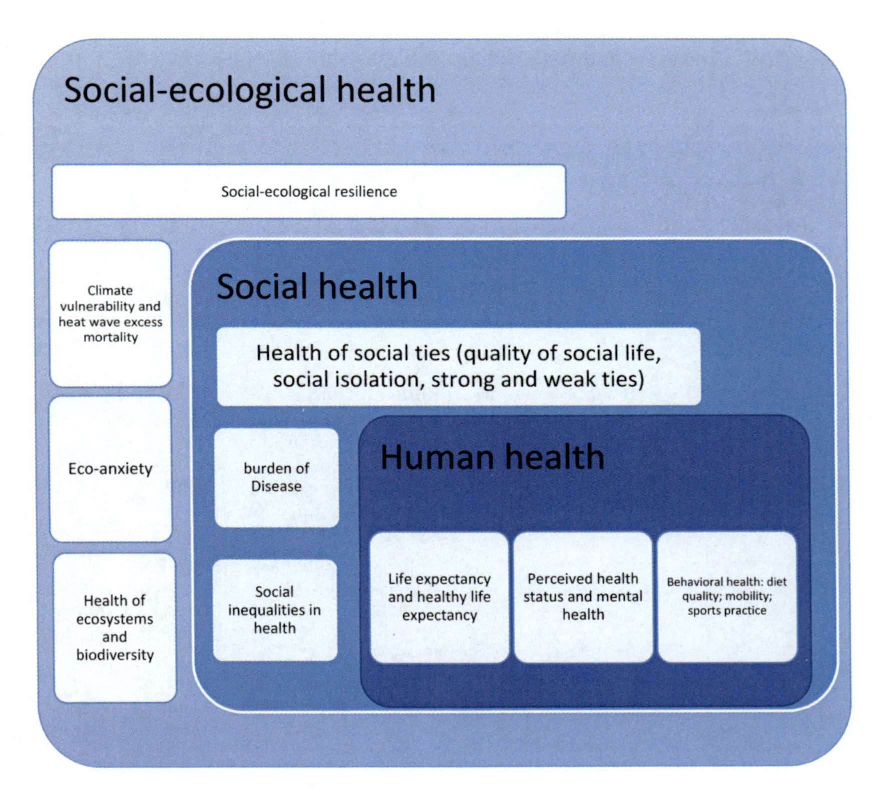

Fig. 3.4 Full health: dimensions and indicators

Table 3.1 Indicators and data of full health

Dimension	*Scope*	*Indicators*	*Source of available data*
Human health (physiological and psychological)	Individual	Life expectancy and life expectancy in good health Perceived health status and mental health Behavioral health: diet quality; mobility; exercise	WHO
Social health	Human community	Burden of disease Health of social ties (quality of social life, social isolation, strong and weak ties) Social inequalities in health	Global Burden of Disease survey CDC Marmot Review (2020)
Social-ecological health	Ecosystem	Climate vulnerability and heat wave excess mortality Eco-anxiety (Hickman et al. 2021) Health of ecosystems and biodiversity Social-ecological resilience	Climate Vulnerability Index (*Lancet Countdown*) *Lancet Countdown;* IPBES World Bank, Resilience rating system

Source Own elaboration

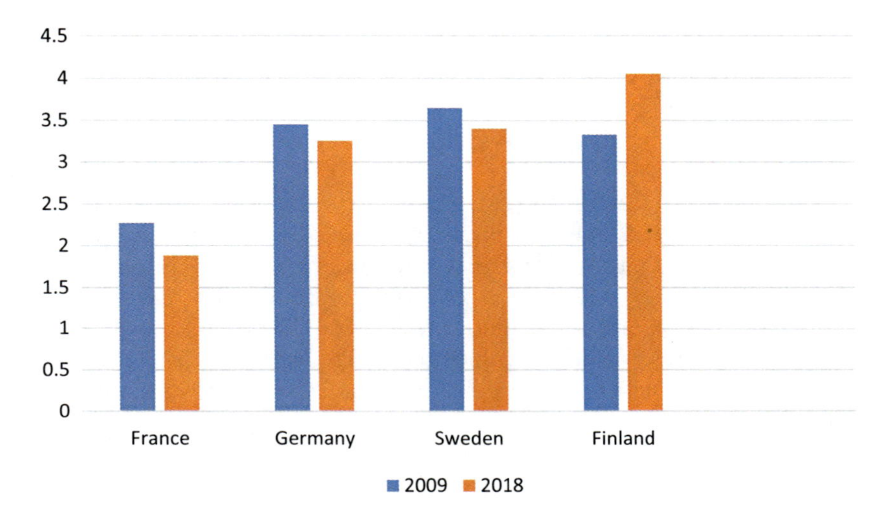

Fig. 3.5 Percentage of total health expenditure devoted to prevention for 2009–2018, selected countries (*Source* Eurostat)

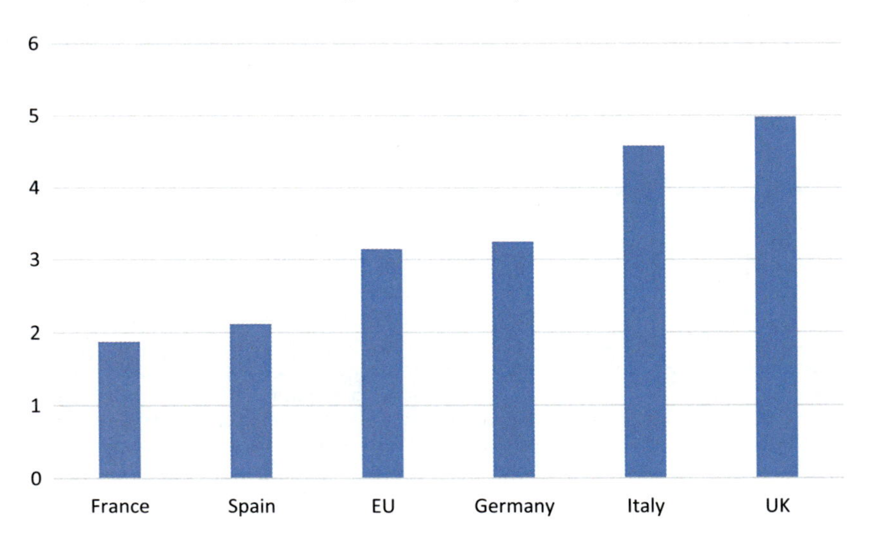

Fig. 3.6 Percentage of total health expenditure devoted to prevention in 2018, selected countries (*Source* Eurostat)

The rate of mortality avoidable by treatment[12] in France is among the best in the world, well below the average for OECD countries as well as that for European Union countries. Conversely, preventable mortality is higher than in many EU countries, although it remains below the OECD average.

The health systems in current growth-dependent welfare states are too oriented towards the treatment of pathologies and the avoidance of mortality by therapy and insufficiently towards the prevention of the burden of morbidity, which, in a context of health fragility, can become a source of financial unsustainability and in a context of ecological shocks can become a cause of health vulnerability. The weakness of prevention and sufficiency policies in the healthcare system is particularly notable and damaging in this respect.

The Covid crisis is in a case in point for the dysfunctional nature of the French healthcare system. The extent of the Covid crisis in France can indeed be understood as a warning: the state of health of the population is marked by significant fragility, growing inequalities and the country's vulnerability to ecological shocks, raising fears of significant human impacts in the coming decades.

France has been one of the hardest-hit countries, especially in 2020 but also in 2021, with a significant excess mortality and a decrease in life expectancy (Laurent et al. 2022). The Covid shock has been wide-ranging in the country: data from the DREES (2021) on the consumption of medical care and goods in 2020 show an increase of 0.4% in value (after

[12] According to the 2019 OECD/Eurostat definitions, preventable mortality is defined as the causes of death that effective public health and primary prevention interventions (i.e., before the onset of disease/injury, to reduce its incidence) would essentially prevent. Treatment-preventable causes of death are those that can be avoided through effective and timely health care, including secondary prevention and treatment interventions (after disease onset, to reduce case fatality). The two current lists of preventable and treatable causes of death were adopted by the OECD and Eurostat in 2019. OECD/Eurostat (2019), "Avoidable mortality: OECD/Eurostat lists of preventable and treatable causes of death", OECD, Paris, http://www.oecd.org/health/health-systems/Avoidable-mortality-2019-Joint-OECD-Eurostat-List-preventable-treatable-causes-of-death.pdf; it is often reported that certain diseases are not "preventable" because there is a genetic or epigenetic predisposition in individuals to these diseases. If this assertion cannot be completely dismissed, the percentage of these diseases which would occur in the absence of any pollution or any other factor (tobacco) is very low. Less than 5% of lung cancers are of genetic origin.

+2.0% in 2019 and +1.5% in 2018), the weakest increase in these expenditures ever observed since 1950, the first available year of health accounts (with an overall 4% increase in healthcare spending due to the pandemic—masks, tests, etc.). This means that the French population has given up on treatment to a massive extent because of Covid, so that the country's chronic disease epidemic will worsen further in the years to come. We see here how more overall spending on therapeutics can result down the road in worse health outcomes. Thus, the mental health crisis in France (partly linked to social devitalization) existed before Covid (increase in depressive syndromes between 2014 and 2019), was aggravated by the restrictions resulting from this crisis (between 2019 and 2020) and has not returned to its state prior to the crisis since the ending of restrictions: the mental health of the French is therefore not resilient to ecological shocks.

A healthcare policy calibrated for the twenty-first century could mean investing in prevention, community building (by mitigating social isolation) and environmental policy to buffer ecological shocks (greening urban spaces, etc.), in other words, focusing the healthcare system on "full health". The potential social savings associated with such a health policy (focused just on mitigating social isolation, air pollution and unhealthy nutrition) vastly exceed these prevention costs and include monetary savings in health expenditure related to the mitigation of avoidable pathologies, non-monetary gains in terms of well-being linked to the prevention of pathologies and social isolation, from life expectancy and disability life expectancy gains to health and well-being benefits, as well as avoiding expenses due to the mitigation of the impact of ecological shocks by preventing and combating social isolation (especially heatwaves but also mental distress linked to lockdowns).

But avoidable mortality can be understood in a much broader sense than the 2019 joint OECD/Eurostat list discussed above: studies of the burden of disease in the United States suggest that almost half of the total morbidity burden is linked to 84 modifiable risk factors and that these preventable diseases are responsible for more than a quarter of total health expenditure (about US$750 billion in 2016). Globally, it is accepted that between 25 and 50% of mortality is avoidable mortality.[13]

[13] Martinez et al. (2020).

US researchers' analysis suggests that high body mass index, high systolic blood pressure, high fasting blood sugar, dietary risks and exposure to tobacco smoke account for the bulk of preventable disease spending.[14] Preventing these risk factors would require a commitment to subsidizing the availability of nutritious foods, discouraging the commercial production of harmful products, investing in early childhood education that leads to healthy exercise and eating habits, and creating cities that encourage healthy behaviors.[15] It would also require a recognition of the role of inequalities of wealth and opportunity in narrowing pathways to better health. All of these would produce a substantial long-term return on investment.[16]

In contrast to these insights, we are witnessing in many developed and emerging countries a particularly counterproductive interplay based on an agri-food system where hyper-packaged products poor in nutrients and harmful to health are heavily subsidized while incurring considerable costs for the healthcare system (80% of healthcare expenditure, which represents 20% of the American national income, is used to treat preventable pathologies such as type 2 diabetes, half of the American population being today either diabetic or pre-diabetic). Rather than increasing the use of medication and/or surgery procedures to treat widespread diseases such as obesity among children,[17] resulting in additional financial burden for healthcare systems, regulating the food industry would be a far more cost-efficient solution.

Just Transitions

The notion of just transition has a particularly interesting history to which it is important to return briefly. It was born in the early 1990s in American trade union circles as a defensive social project aimed at protecting workers in the fossil-fuel industries from the consequences of

[14] The US Burden of Disease Collaborators (2018).

[15] Martinez et al. (2020).

[16] Bolnick et al. (2020).

[17] "American Academy of Pediatrics Issues Its First Comprehensive Guideline on Evaluating, Treating Children and Adolescents With Obesity", September 1, 2023 https://www.aap.org/en/news-room/news-releases/aap/2022/american-academy-of-pediatrics-issues-its-first-comprehensive-guideline-on-evaluating-treating-children-and-adolescents-with-obesity/.

climate policies for their jobs and their pensions. For US trade unionist Tony Mazzocchi (Mazzochi 1993), just transition meant protecting jobs, wages and pensions from the effect of environmental policies (e.g., environmental standards regulating the fossil-fuel industries). But Mazzocchi was also instrumental in the passage of the 1970 Williams–Steiger Occupational Safety and Health Act, linking employment with environmental conditions and contributing to establishing health as the cornerstone of human well-being.

There are indeed two imperatives in the just transition that can be seen, at first glance, as competing with one another. The first principle is the willingness to protect jobs and pensions from the effect of environmental policies (e.g., environmental standards), which was the initial focus of the just transition when it was formulated three decades ago as a social compensation policy (Mazzochi 1993); the second principle is the willingness to protect the health of workers, communities and ecosystems from pollution and more generally environmental degradations induced by income-generating economic activity.

The first principle found a contemporary echo in the EU with the Silesia/Katowice Declaration in 2018 and the creation of the "Just Transition Mechanism" of the European Green Deal in 2019. At the global level, it was taken up in the Paris Agreement of 2015 (which refers to the "imperatives of a just transition for the active population and the creation of decent and quality jobs in accordance with the development priorities defined at the national level"). During COP 26, on November 4, 2021, several heads of state and government (including those of the co-organizers Italy and the United Kingdom, but also of France, the European Commission and the United States) co-signed a "Declaration on International Just Transition".

From this defensive perspective (which we find in the current debates in the United States around the future of coal-producing states like West Virginia), it is the transition policies that must be made just. However, the amplification of ecological shocks (floods, droughts, pandemics, etc.) in all corners of the planet, independently of the mitigation policies that will be implemented to deal with them, calls for a broader and more positive definition of the just transition.

This enlargement was initiated under the influence of the International Trade Union Confederation and then by the European Trade Union Confederation, which made the just transition evolve towards an attempt to reconcile the fight against climate change and the reduction of social inequalities, around the theme of "green jobs" and the slogan "No jobs on a dead planet". This social-ecological project is explicit in the 2016 report of the International Labor Organization which defines "guidelines" in this area (ILO 2016).

It is this broader definition that we find in the Declaration of November 4, 2021, which takes up the traditional themes of supporting workers in the transition to new jobs characterized by decent work through social dialogue, but embedded in a new economic strategy which notably involves redefining growth models considered unsustainable on the ecological (overconsumption of resources) and social (exacerbation of inequalities) levels.

If this position is welcome, it is still insufficient: it is necessary to further broaden the just transition project by specifying its requirements and above all by striving to make it operational in a democratic way.

The just transition must no longer be understood only as social support or financial compensation for policies aimed at mitigating ecological crises, but more broadly as an integrated social-ecological transition strategy in the face of ecological crises, including both ecological policies and environmental shocks (a carbon tax is an ecological policy while a heat wave is an ecological shock).[18]

The COVID-19 crisis is a good illustration of the relevance and necessity of this broader just transition: it is an ecological shock (in this case, a zoonosis) which has aggravated existing social inequalities (housing, essential workers, comorbidities, etc.) and has given rise to new ones (need for/possibility of remote working, long Covid, etc.). In this spirit, three requirements of a just transition strategy can be defined:

1. Systematically analyze ecological shocks and the policies that intend to mitigate them from the perspective of the three fundamental dimensions of social justice: recognition, distributional and procedural;
2. Prioritize in the design of just transition policies social-ecological well-being informed by these issues of justice in order to go beyond the horizon of economic growth;
3. Build and implement these just transition policies in a democratic way by ensuring the understanding, support and commitment

[18] See Bauler et al. (2021).

of citizens, at the different levels of government (local, national, European and global).

To detail this analytical framework, let us start from a widely shared point of view: the most disadvantaged would be the big losers of ecological transition policies, and this social reality will only get worse as these policies gain momentum. It is the core misinterpretation that has marred the analysis of the "yellow vests" crisis in France (2018–2019): energy-climate transition gives rise to inevitable social inequalities that exasperate citizens. In reality, current social inequalities are largely the result of the existing economic system and non-transition policies. There is no clearer illustration of the cost of non-transition than the surge in inflation that began in the fall of 2021, and which is the product of oil and gas dependence, fuel poverty and geopolitical volatility immediately translating into social vulnerability. In reality, just transition policies are both feasible, inexpensive and potentially acceptable to a majority of citizens. These just transition policies can take three forms:

1. Mitigating the ecological non-transition, i.e., the current situation in which ecological crises generate social inequalities which affect first and foremost the most deprived;
2. Reducing social inequalities can mitigate ecological crises, and reciprocally, ecological transition policies can reduce social inequalities and improve the well-being of the poorest;
3. Designing social-ecological policies which, now and in the long term, simultaneously reduce social inequalities and environmental degradation.

In relation to the first type of policies, different categories of environmental inequality exist and must be broken down to be properly identified and possibly addressed and mitigated. A first typology of the generative factor of environmental inequalities (the event generating the inequality) consists in dividing them into two categories: the inequality impact *of* individuals and groups on environmental damage and definition of environmental policies; and the inequality impact *on* individuals and groups of policies and environmental damage. A second typology of environmental inequalities consists in considering their inequality vector: what form of environmental degradation is responsible for the observed injustice. A third typology looks at criteria of inequality: what dimension of human beings is at play in the observed injustice. Synthesizing these criteria into a framework, four types of environmental equality thus appear (Table 3.2):

- Type 1 is concerned with procedural justice and stems from the potential exclusion of individuals and groups from public policy procedures, for instance the inability to participate in decision making on the installation of polluting sites in their residential area.
- Type 2 is concerned with recognition justice and stems from the potentially adverse social effects of environmental policy on individuals and groups.
- Type 3 is concerned with distributive justice and stems from the unequal exposure and sensitivity of individuals and groups to environmental degradation, for instance, the heavier pollution burden placed on disadvantaged neighborhoods in metropolitan areas.
- Type 4 is also concerned with distributional justice but stems from the unequal responsibility of individuals and groups in environmental degradation, for instance the greater pollution footprint of richer households.

By combining these elements, it can be analytically assessed that the environmental inequality experienced by a Parisian child living near dense traffic during a spike of pollution due to PM2.5 is an inequality of exposure whose vector is air pollution and the criteria are age, neighborhood and locality (at play with possible others such as ethnicity and income level).

A second aspect of just transition policy is the reduction of social inequalities that have increased sharply in recent decades , and there is every reason to want to reduce them for social and economic reasons. The just transition must begin with the most affluent and powerful—who are both more responsible for our ecological crises and more able to contribute to their mitigation—so that everyone can be called upon to contribute and evolve. It turns out that this reduction can generate an additional beneficial effect: the alleviation of ecological crises (this is the converse of the reasoning described in the previous chapter, which detailed how social inequalities degrade the environment).

Consider the fiscal tool. We may want to restore some of the social justice lost since the early 1980s by increasing the level of levies on

Table 3.2 A typology of environmental inequality

	Philosophical approach	*Generative fact*	*Inequality vectors*	*Inequality criteria*	*Environmental inequality examples*
Types of air inequality					
Type 1	Procedural justice	Impact of individuals and groups on environmental policies	Exclusion from public decision-making procedures	Nationality, spatial location, age, gender, socio-economic level (income, health, education, etc.), ethnic characteristics, etc	Non-participation in the decision to install a toxic site (e.g., a chemical plant) in the city of residence
Type 2	Recognition justice*	Impact of environmental policies on individuals and groups	Taxation, regulatory policies, information/ awareness		Vertical and horizontal income inequalities caused by banning polluting vehicles
Type 3	Distributive justice	Exposure / sensitivity (vulnerability) to environmental degradation	PM 2.5; PM 10		Unequal exposure and sensitivity to air pollution in urban areas
Type 4	Distributive justice	Impact of individuals and groups on environmental degradation and resource consumption	NO_2; NOx; SO_2; O_3		Air pollution by the top income deciles

Source adapted from Laurent (2022)

*This is a process model of social justice that includes a positive regard for social difference and the centrality of socially democratic processes companies (especially in the digital sector) and the highest incomes (especially millionaires and billionaires, who have never been so numerous). But we can calibrate this taxation to maximize its ecological impact, for example by using the two tax bases that are wealth and CO_2 consumption.

By taxing wealth, we can tax past unequal growth without the need for additional growth, but we can also tax income, i.e., implement a social-ecological taxation of inequalities in income and carbon footprint, provided that we start from a lucid diagnosis of the state of inequalities. Environmental taxation is indeed a case in point of an ecological policy that can lead to aggravating injustices by claiming to want to correct them. But designing and implementing just transition policies is simple, inexpensive and independent of growth. Progressive social-ecological taxation is simply defined as targeting the carbon footprint of the upper deciles, the revenue collected being used to finance the reduction of emissions of the lower deciles, via public investments (for example in the renovation of housing): a tax and transfer policy is able to connect reductions in "luxury emissions" (air and road leisure travel, luxury consumption) to reductions in "essential emissions" (food, housing and work mobility).

It can be considered fair if it is based on a fair contribution in terms of resources and contributes to satisfying energy needs to enable everyone to live with dignity. These two conditions can easily be translated into social compensation measures that would lead to the redistribution to a majority of French people of the purchasing power drawn from the receipts of this progressive social-ecological taxation, and to significantly and durably reducing fuel poverty (Berry and Laurent 2019), which was hard felt in France (as in the rest of Europe) during the winter of 2021–2022.

Many options or methods of this progressive social-ecological taxation exist. In the case of France, many progressive social-ecological taxation options exist. For instance, by increasing the currently frozen carbon tax to €55 per ton of carbon as an environmental objective and redistributing 25% of revenues to households using already existing mechanisms, a majority of households (more than 50% of households in the lowest six income deciles) could gain from carbon taxation (receiving more in social transfers than what they pay in carbon taxation). The remaning 75% of the revenue could be allocated to mitigating fuel poverty but also to providing financial help to shift to low-carbon equipment, reducing social inequality further in a context of rising energy prices (Berry and Laurent

2019). Hence, progressive social-ecological tax policies may be able to both lower the carbon footprint of the highest deciles while redistributing money to compensate the reduction of lower deciles while allowing them to invest in low-carbon lifestyles. These strategies for "recycling" revenues from environmental taxation are at the heart of successful experiences of carbon taxation in Europe and around the world and have been identified as the key to the acceptance of these taxes by populations (see Andersson and Atkinson 2020 for application to a broad set of countries). We can also drastically reduce fossil-fuel subsidies that benefit large companies to free up considerable resources for the ecological transition (fossil-fuel subsidies have never in history been higher than in 2022).[19]

We can extend this logic of social compensation to non-monetary gains in well-being, starting with the health benefits generated by ecological transition policies. The European Environment Agency (European Environment Agency, 2021) has recently proposed in this spirit both to "minimize the monetary inequalities resulting from the transition to carbon neutral economies" and to "maximize the non-monetary co-benefits, such as health co-benefits" (on this notion of co-benefits, see Chapter 4). It is essential to include these non-monetary gains from environmental policies in the analysis because they mainly benefit the most disadvantaged (Drupp et al. 2018).

Such measures are the perfect segue into the third and last type of just transition policies: integrated and sustainable social-ecological policies, which go beyond the sole logic of social compensation for the ecological transition.

Implementing a just transition policy means articulating social issues and environmental challenges to allow progress simultaneously in both dimensions, either because progress in one leads to progress in the other (as in the case of the thermal renovation of a dwelling, where environmental progress leads to social progress), or because the result of the social-ecological policy is parallel progress in both areas. But, in many cases, envisaging and designing a just transition policy consists in recognizing the need for arbitration between the social question and the

[19] According to the IEA, "In 2022, subsidies worldwide for fossil fuel consumption skyrocketed to more than USD 1 trillion, according to the IEA's latest estimate, by far the largest annual value ever seen" with subsidies for natural gas and electricity consumption more than doubled compared with 2021, while oil subsidies rose by around 85% (IEA 2023).

environmental question (an example being carbon taxation, which can, if we are not careful, have harmful social consequences).

A recent study carried out among more than 40,000 respondents in 20 countries representing 72% of global CO_2 emissions allows us to understand precisely how a just transition policy can make an environmental policy possible, in this case climate policy (Dechezleprêtre et al. 2022). The authors show first, that in all countries, support for climate policies depends on three key factors—the perceived effectiveness of policies in terms of reducing emissions, their perceived distributional effects on low-income households (inequality concerns) and their own household's gains and losses—and second, that the key to political support for environmental policies is to explain how policies work and who can benefit from them, which is essential for fostering political support. Thus, when a tax of €45 per tonne on fossil fuels is submitted to respondents (an ecologically effective measure to combat climate change but socially unjust—see Fig. 3.7), it does not receive a majority anywhere, but when this is accompanied by compensation for the poorest households, it becomes the majority (sometimes strongly, in five major OECD economies out of eight—see Fig. 3.7).

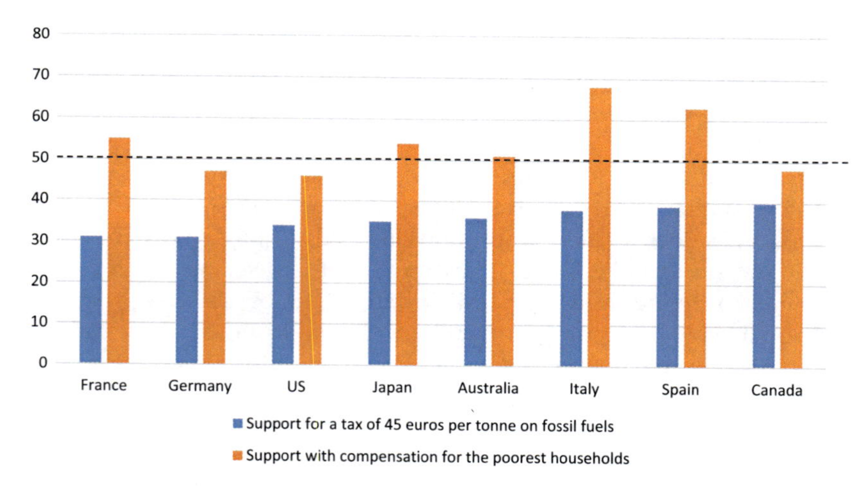

Fig. 3.7 Support for carbon taxation, without and with social compensation (% of respondents) [*Source* Dechezleprêtre et al. (2022)]

A final, critically important dimension of the just transition is its democratic nature: it is here that the "procedural" dimension of just transition policies comes into play, the just transition being both a goal and a method to reach this goal. Justice is a condition of the transition, that is to say both a lever to activate attitudes and behaviors and an end in itself. The challenge of ecological transition thus invites us to imagine new forms of social cooperation but also to strengthen existing forms of social cooperation by developing, for example, methods known as "open democracy", such as the Citizens' Climate Convention, launched in France in October 2019.

In the wake of the "Yellow Vests" crisis, a collective called the "Citizens' Vests" (led by the director and poet Cyril Dion and the actor Marion Cotillard) made an offer to President Emmanuel Macron in January 2019 to organize a "citizens' assembly" to debate in particular the social issues raised by the ecological transition. The proposal was accepted in the spring and 150 representative citizens (randomly selected according to gender, age, level of diploma, socio-professional category, type of territory and geographical area) were collectively entrusted with the following mandate: "Define structural measures to achieve, in a spirit of social justice, a reduction of GHG emissions by at least 40% by 2030 compared to 1990". On the basis of shared knowledge and trust, cooperation between the members of the Convention has enabled 149 of the 150 proposed to be unanimously approved, the quality of the work carried out being widely praised by both experts and environmental organizations and associations (Giraudet et al. 2020).

While the ecological transition is supposed to be made impossible by the divorce between social classes whose lifestyles and cultural reflexes have become antagonistic and irreconcilable, it is actually possible for very different people to engage in dialogue and exchange in depth—in the light of the most advanced scientific and technical knowledge—about public policies for climate transition.[20]

[20] In addition to the quality of the work accomplished in a short period of time by a citizens' assembly, the orientation of the measures recommended by the Convention is proving rich in lessons for just transition policies. The members of the Convention considered that social justice would be served more by regulation and standards than by taxation: of the 149 measures finally adopted, 76 were of a regulatory nature and only eleven of a fiscal nature (The remaining 71 measures do not fall into either of these two categories, either because they are general in scope or because they fall into another category of public instruments, such as awareness-raising measures).

In addition to the quality of the work accomplished in a short period of time by a citizens' assembly, the orientation of the measures recommended by the Convention is proving rich in lessons for just transition policies. The members of the Convention considered that social justice would be served more by regulation and standards than by taxation: of the 149 measures finally adopted, 76 were of a regulatory nature and only eleven of a fiscal nature (the remaining measures do not fall into either of these two categories, either because they are general in scope or because they fall into another category of public instruments, such as awareness-raising measures).[21]

Social-Ecological Regeneration

There is a direct link between our unique capacity for social cooperation and the need for ecological transition: in the light of Elinor Ostrom's pioneer work, we know that common institutions rooted in principles of justice, even when reduced to their simplest expression, favor the cooperative behaviors likely to perpetuate the human communities that welcome them. The reinvention of social innovation under ecological constraint enables regeneration of social cooperation which in turn regenerates the living world through non-destructive forms of social cooperation. The living world and social world regeneration processes go hand in hand: by relying on what sets us apart in the living world (social cooperation), we can find our way back to our proper place within the living world.

The work of Ostrom (Ostrom 1990, 2010, 2012) and her many co-authors indeed demonstrates that the institutions that allow the preservation of resources through cooperation are generated by human communities themselves, and not by the state or by the market. It is therefore a double invalidation of Hardin's hypothesis of the "tragedy of the commons" (cf. supra): cooperation is possible and it is self-determined. Hundreds of decentralized governance systems have been avoiding the tragedy of the commons all over the world for millennia by allowing the sustainable exploitation of all kinds of resources: water, forests, fish,

[21] In his speech of April 25, 2019, President of the Republic Emmanuel Macron promised to submit directly and in their integrity ("without filter") the measures proposed via referendum, vote of Parliament or direct regulatory application, but he did not live up to his promise, which does not prove that open democracy methods do not work, but rather that French representative democracy is dysfunctional.

etc. This has been the case in particular with the sharing of water since the beginnings of human agriculture 10,000 years ago. These commons involve a well-defined community of users, as well as a set of rules and norms that allow everyone to regulate the behavior of others as well as their own, so they are not natural resources but social-ecological relations.

Let us insist on this essential point: these principles of ecological government emanate from human communities themselves, not from an external authority, even if the latter can serve as guarantor of these rules. Conversely, authoritarian constraints imposed on local groups by distant governments often turn out to be counterproductive, because the authorities in question do not have sufficient information or legitimacy to act in the long term (the solutions for the privatization of natural resources also suffer from many limitations, starting with the inequalities they induce).

Ostrom starts from a fundamental discovery made in the laboratory by means of "games": individuals cooperate much more than the standard theory presupposes.

In the so-called "public good" game, individuals cooperate much more than the standard theory presupposes, especially if they have the opportunity to punish the "free riders" dear to Olson. This game is based on a somewhat complex set-up. The organizers explain to the participants that, during several consecutive rounds, they will be associated with other players anonymously, each one being endowed with a starting bet, say €20. In each round, all players are faced with two options: either place their money in a common account, or deposit it in a private account, knowing that the winnings from the common account are returned to each private account at the end of each round on a pro-rata basis according to their contributions. However, the return of the joint account is lower than that of the private account, so that its profitability depends on the cooperative goodwill of the players (the higher the contributions to the joint account, the more it will bring to each player). The profitability results of the joint account are announced at the end of each round of play, which allows each player to guess the decisions of the others and to adapt theirs accordingly in the next round.

In such a game, the logic of individual interest and uncertainty about the cooperative capacity of other players should induce individuals not to cooperate, in other words never to contribute to the common account. However, in reality, the results are quite different: the players begin by investing on average half of their endowment in the common account. In other words, the willingness to cooperate is much stronger than expected, especially if players are given the means to punish secessionists (for example via financial fines imposed on those who never contribute).

However, cooperation can gradually wither away. After the first cooperative impulse, we observe a decrease in the average contribution levels between the first and the last period of the game, when the game is repeated. The players converge towards contributions which tend towards 0 after 10 to 15 periods, and, soon, no one contributes to the public good anymore, because everyone is convinced that the others will default.

From this result, we can distinguish three categories of players: altruists, conditional cooperators and stowaways. Altruists cooperate all the time and free riders never, while conditional, wait-and-see cooperators begin to cooperate but then cease due to disappointment or even resentment of the free riders. This non-cooperative spiral can easily be illustrated: if fraud in public transport, which currently accounts for around 2% to 3% of passengers in a city like Paris, were to become widespread, a majority of users would end up refusing to paying for their tickets, making it impossible to operate and maintain the network. Empirical work has established that 30% of gamblers behave like free riders, 50% are conditional cooperators and 20% altruists.[22] If the 30% of free riders convince the 50% of conditional cooperators not to cooperate anymore, cooperation is over.

When punishments are allowed, cooperation does not deteriorate as the game progresses. The game of public good therefore shows individuals who cooperate spontaneously, who may be discouraged, even disgusted, by the sabotage of free riders, but who may also regain faith in cooperation if non-cooperative behavior is sanctioned.[23]

Ostrom verified her intuition of a conditional inclination toward cooperation in the field around the world: in hundreds of meticulously documented cases, humans manage to avoid the "tragedy of the commons" by building collective rules whose pillars are reciprocity, trust and justice. Whether it is rivers to be preserved from pollution, forests that must be exploited reasonably while maintaining them, fish that must be caught in moderation to enable them to reproduce, from Switzerland to Japan, Spanish irrigation systems to Nepalese irrigation systems, humans show themselves capable of cooperating to preserve, conserve and thrive. Based

[22] Fischbacher et al. (2001).

[23] Fehr and Gächter (2000).

on her field observations, Ostrom endeavors to define the main principles of effective management of common resources in order to overcome what she calls the "social dilemmas" of environmental governance. These principles can be simply understood as the rules of the game of human cooperation.[24]

According to Ostrom, individuals who cooperate are able to learn from others; they remember the cooperative behaviors and, more generally, the reliability of the people they have dealt with; they use their memory and other cues (such as language or general attitude) to assess the reliability of their partners in the exchange before granting them their trust; they strive to build a reputation for reliability; they punish the "free riders", even if it incurs a cost for them too; and they adopt time horizons that go beyond the immediate past. These characteristics can only develop through the use of the cognitive faculties of individuals within the framework of a process of socialization. In other words, cooperation is indeed a quest for shared knowledge.

Laboratory experiments, field work, empirical devices, theoretical framework: Ostrom juggled, hard at work in her workshop at Indiana University, with methods and approaches, between political science, social psychology and environmental studies to renew in depth the discipline of economics and to transmit to us a tremendous lesson of hope regarding the continuation of the human adventure on the planet. Yes, human collective intelligence can do anything, provided we understand that the technology of the future in which we excel is social innovation.

Faced with each new challenge, human communities reinvent the institutions of cooperation based on principles of justice, which leads to the common good. Through what appears at first sight to be a collaboration limited to a precise object (the preservation of a lake or an animal species), human communities discover the way to govern themselves according to rules which they consider as fair. The fatalism of Hardin, who advocated that human freedom itself was at the root of the failure of cooperation,

[24] Namely: 1. Define clear group boundaries; 2. Match rules governing use of common goods to local needs and conditions; 3. Ensure that those affected by the rules can participate in modifying the rules; 4. Make sure the rule-making rights of community members are respected by outside authorities; 5. Develop a system, carried out by community members, for monitoring members' behavior; 6. Use graduated sanctions for rule violators; 7. Provide accessible, low-cost means for dispute resolution; 8. Build responsibility for governing the common resource in nested tiers from the lowest level up to the entire interconnected system (Ostrom 2011).

is completely reversed. Thanks to Ostrom, we now know that common institutions rooted in the principles of justice, even when reduced to their simplest expression, promote cooperative behavior. Ostrom's theory of the commons therefore constitutes the matrix of social-ecological cooperation. The benefits yielded by the power of social cooperation extend to natural systems regeneration.

Yet the dominant approach today is a "separatist approach": it aims not only to erect an impermeable border between social and natural systems but to divide and isolate the natural systems themselves by instrumentalizing them, reifying them and monetizing them in order to use them to feed a new form of economic growth, sometimes called "green growth", but which it seems more accurate to call "bio-growth": the growth of the gross domestic product (GDP) drawn from the exploitation of the living.

This approach, which has its own internal consistency, responds to two intertwined logics. On the one hand, it is a matter of taking note of the domination of economic imperatives (income, profit, growth) over natural and social dynamics in order to generalize them to all human interactions, including non-market transactions, by far the more numerous. On the other hand, within this framework, it is important to seek practical solutions to the problem supposed to be the keystone of the biodiversity crisis: the value of nature. The separatist approach is thus based on a triptych: separating the living world from the social world by making economic growth a compass, petrifying the living world by means of "natural capital" and fragmenting it by pricing non-human species on the pretext of preservation (by treating the living as an externality).

The Dasgupta Review on the economics of biodiversity commissioned by the UK government (Dasgupta 2021), for instance, proclaims in its first pages that the biodiversity crisis results from a lack of optimality in the "portfolio management of natural assets" for which humanity is responsible. "The three ubiquitous features of the natural world—mobility, silence and invisibility—prevent markets from adequately recording our use of nature's goods and services. "It is therefore necessary, according to the Review, to remedy this with the help of economic instruments which can lead to prices fixing the economic information that the markets will then make public. Nature, returned to its economic vocation, will then be motionless, noisy and tangible.

The alternative path I propose here to follow, along with many others, here diverges radically from this separatist approach. It consists of a holistic approach to social and natural systems that puts the economy back in its proper place and is based on an intuitive notion: bio-solidarity. Because if the term biodiversity, invented in the mid-1980s, captures the variability of life on Earth, it does not translate well its essential principle: the interdependence of life forms. Because the living world is not only an aggregate of flows, but even more a set of links—it is a dynamic network of natural and social relations—today in danger.

Bio-solidarity thus designates the interdependence of the different forms of the living world, including humans, and is measured by the vitality of the links that unite both non-human species with each other and these species with the human species. The key indicator of vitality is not only variety but also togetherness (Laurent and Morand 2021). Bio-solidarity therefore aims not to count or inventory species but to recognize and identify natural and social links. The holistic approach to the preservation of biodiversity therefore makes health the great mediator between species and underlines the role of social ties in this "full health", ties patiently woven over centuries by Indigenous communities around the world (FAO 2021).

Ecosystems and biodiversity are not overexploited for lack of economic value but because of economic value. The problem, in other words, is not the value of nature, but the nature of value. A recent study by the Intergovernmental Science-Policy Platform on Biodiversity and Ecosystem Services (IPBES) shows that, while there are currently around 50 methods for assessing the different values of natural resources (cultural, intrinsic, etc.), three-quarters of the existing studies focus solely on their "instrumental" value and neglect their relational value (IPBES 2022). Ecosystems and the biodiversity that underpins them do not suffer from a lack of valuation but from a conflict of values that today turns to the advantage of economic value to the detriment of all others. Because the missing value hypothesis is as dubious as it is debatable and can easily be reversed. The economic value of natural resources is in fact very well understood and assimilated by those who benefit from them, which is precisely the reason for the power asymmetries that are exercised to appropriate them (see on this point the line of analysis developed in the last 30 years by James Boyce).[25]

In its overall assessment, the IPBES (IPBES 2019) underlines the fact that the degradation of biodiversity and ecosystems is generally less

[25] Boyce, 1994; Boyce, 2002; Boyce, 2019.

rapid in territories managed by Indigenous peoples than in others. This is particularly true of a strategic territory of the biosphere, the Amazon rainforest, where deforestation has been visibly less tangible for 20 years on lands controlled by Indigenous communities.[26] A recent report offered a comparison between existing territorial management regimes and notes:

> Protected Areas (PA) and Indigenous Territories (IT) are vital to protect the Amazonia. Between both regimes (PA and IT), about half (48%) of the Amazon is covered; however, the other half (52%) are undesignated areas that are in danger of disappearing and without which it is impossible to avert the tipping point. Most of the deforestation (86%) took place outside national PA and IT ... 255 million hectares of intact areas and Key Priority Areas with low degradation have not been titled to indigenous peoples or designated as protected areas and are at imminent risk. The undesignated areas register the greatest transformation (33%) and high degradation(10%) being six times more the transformation registered in the PAs and more than eight times that of the IT (Quintanilla et al. 2022).

But this observation is even more general: many works show that "equitable conservation", which fully recognizes the role of Indigenous peoples and Indigenous communities, is the most promising way to preserve biodiversity and ecosystems. Conversely, externally imposed approaches to conservation that aim to protect natural resources by ignoring the social-ecological relationships that have developed between humans and other species are not only unfair but above all ineffective (Dawson et al. 2021; FAO et al. 2021). Thus the "30 by 30" project, aimed at conserving 30% of land and sea surfaces in 2030 and now supported by the United Nations, is the subject of criticism from Indigenous movements, who fear mass expulsions in the territories concerned in the name of protecting biodiversity.

Indigenous environmentalism can certainly lay claim to the title of the oldest tradition of environmental justice, but it is also a tradition that is not explicit in the very communities where it has been practiced

[26] A recent report notes: "For millennia, through ancestral practices and knowledge, the Amazonian indigenous peoples and nationalities have protected our forest and all the life that nests in its trees and that flows through the rivers of the largest basin on the planet" ... "We know that intact ecosystems and low degradation areas represent 74% of the Amazon and that we can still restore 6% to achieve the protection of 80% of the region and avert the tipping point." (Quintanilla et al. 2022).

until today. The assumption that Indigenous peoples were the "original ecologists" or "original conservators", living in perfect harmony with the environment, is doubly wrong: it caricatures the natural world as harmonious and Indigenous peoples as "noble savages". Far from a nativist romanticism, the observation of a certain coherence is unfortunately essential: the destruction of ecosystems has been accompanied by the destruction of the human communities that take the best care of them.

It is thus undeniable that the ecological crises bear the mark of a brutal colonialism which led to the appropriation of natural resources by the Western powers, to the destruction of many Indigenous communities living from these resources and consequently to the destruction of social-ecological relationships that these communities had been able to weave over the centuries. As Hickel (2021) notes, "The economic growth of the North rests on the patterns of colonization: the appropriation of the atmospheric commons, the appropriation of natural resources and labor from the South. Both in terms of emissions and resource use, the global ecological crisis runs in the colonial line." Conversely, as the British historian Richard Grove (1995) argues, it was in the colonies that the first environmental policies emerged under the pressure of economic imperialism and its depredations.

But Indigenous environmentalism does exist, with its complexity and its limits, and it has been rediscovered in the contemporary period by ecological thinkers who have looked to Indigenous peoples for inspiration and good practices in order to reinvent a more balanced relationship between contemporary societies and ecosystems, some of which are hundreds of millions of years old. Indigenous environmentalism stems from an undeniable fact: Indigenous peoples (about 370 million people occupying 20% of the Earth's territory belonging to no less than 5000 different cultures) are more exposed to environmental degradation due to their proximity to the system's natural resources and their social vulnerability.

Recognition of the rights of these Indigenous communities is in no way due to an ecological romanticism according to which noble savages live in harmony with nature while cynical Westerners have forgotten its precepts and benefits: Indigenous peoples contribute to the improvement and maintenance of biodiversity and wild and domesticated landscapes as they practice agricultural methods adapted to local conditions and compatible with the conservation of biodiversity by creating habitats rich in species and of a great diversity of ecosystems in cultural landscapes (grasslands mowing, central Europe) and by identifying useful plants and their cultivation in highly diversified ecosystems (garden forests, Indonesia). The ecological stakes are considerable: more than a third of the forests still intact on the planet—crucial for the preservation of the climate and biodiversity—are located on the lands of Indigenous peoples (Fa et al. 2020).

However, this network of social-ecological relations, which is a reserve of collective intelligence, is also an endangered species: the natural spaces managed by Indigenous peoples under various regimes face growing extraction of resources, mining, energy and transport infrastructure, with various impacts on the livelihoods and health of local communities. These communities, because they are highly dependent on nature and its contributions for their sustenance, health and existence, will be disproportionately affected by these alterations (IPBES, 2019), but they also act as global ecological sentinels. Hence the importance of giving full consideration to Indigenous communities in the Convention on Biological Diversity currently being negotiated at the global level. As David Schlosberg and David Carruthers (2010) argue, "Threats to Indigenous peoples—their rights, lands and cultures—have been a powerful catalyst for mobilization, as Indigenous communities struggle against corporations, governments, policies and other forces that threaten to fragment, displace, assimilate or drive them to cultural disintegration".

Understanding, in line with Ostrom's work, that the living world is not a stock but a set of flows and that natural resources are social-ecological relations, leads to wanting to evolve the reflection of biodiversity towards bio-solidarity as the cornerstone of social-ecological regeneration.

References

Andersson, J., and Atkinson, G. (2020), "The Distributional Effects of a Carbon Tax: The Role of Income Inequality". Centre for Climate Change Economics and Policy Working Paper 378/Grantham Research Institute on Climate Change and the Environment Working Paper 349. London: London School of Economics and Political Science.

Bauler, T. et al. (2021), « La transition juste en Europe : mesurer pour évoluer », *Cahier de prospective de l'IWEPS*, no. 6.

Berry, A., and Laurent, E. (2019), « Taxe carbone, le retour, à quelles conditions ? », OFCE Working Paper no. 6, April.

Bolnick, H. J., Bui, A. L., Bulchis, A. et al. (2020), "Health-Care Spending Attributable to Modifiable Risk Factors in the USA: An Economic Attribution Analysis", *Lancet Public Health*, vol. 5, pp. e525–e535.

Boyce, J. K. (1994), "Inequality as a Cause of Environmental Degradation", *Ecological Economics*, vol. 11, no. 3, pp. 169–178.

———. (2002), *The Political Economy of the Environment*. Cheltenham: Edward Elgar.

———. (2019), *Economics for People and the Planet. Inequality in the Era of Climate Change*. New York: Anthem Press.

Credoc, «10 ans d'observation de l'isolement relationnel: un phénomène en forte progression », Les Solitudes en France.

CNAM. (2021, July), Propositions de l'Assurance maladie pour 2022.

Dasgupta, P. (2021), *The Economics of Biodiversity: The Dasgupta Review*. London: HM Treasury.

Dawson, N. M. et al., (2021), "The role of Indigenous peoples and local communities in effective and equitable conservation", *Ecology and Society*, vol. 26, no 3, art. 19.

Dechezleprêtre, A. et al. (2022), "Fighting Climate Change: International Attitudes Toward Climate Policies", OECD Economics Department Working Papers, No. 1714, OECD Publishing, Paris. https://doi.org/10.1787/3406f29a-en.

DREES. (2021), « Les dépenses de santé en 2020—Résultats des comptes de la santé édition 2021. »

Drupp, M. A., Meya, J. N., Baumgärtner, S., and Quaas, M. F. "Economic Inequality and the Value of Nature", *Ecological Economics*, vol. 150, pp. 340–345.

Ellis, E. C. et al. (2010), "Anthropogenic Transformation of the Biomes, 1700 to 2000", *Global Ecology and Biogeography*, vol. 19, no. 5, pp. 589–606.

European Environment Agency. (2021), "Exploring the Social Challenges of Low-Carbon Energy Policies in Europe", Briefing, no. 11.

Fa, J. E. et al. (2020), "Importance of Indigenous Peoples' Lands for the Conservation of Intact Forest Landscapes", *Frontiers in Ecology and the Environment,* vol. 18, no. 3, pp. 135–140.

FAO, Alliance of Bioversity International and CIAT. (2021), "Indigenous Peoples' Food Systems. Insights on Sustainability and Resilience from the Front Line of Climate Change", Report. Rome.

Fehr, E., and Gächter, S. (2000), "Cooperation and Punishment", *American Economic Review*, vol. 90, no. 4, pp. 980–994.

Fischbacher, U., Gächter, S., and Fehr, E. (2001), "Are People Conditionally Cooperative? Evidence from a Public Goods Experiment", *Economics Letters*, vol. 71, no. 3, pp. 397–404.

Frumkin, H., Bratman, G. N., Jo Breslow, S., Cochran, B., Kahn Jr, P. H., Lawler, J. J., Levin, P. S., Tandon, P. S., Varanasi, U., Wolf, K. L., and Wood, S. A. (2017), "Nature Contact and Human Health: A Research Agenda", *Environmental Health Perspectives*, vol. 125, no. 7.

Giraudet, L.-G. et al. (2020, October 7), "Deliberating on Climate Action: Insights from the French Citizens' Convention for Climate", HAL.

Grove R. H. (1995), "Green Imperialism. Colonial Expansion, Tropical Island Edens and the Origins of Environmentalism", pp. 1600–1860, Cambridge University Press.

J. F. Helliwell, R. Layard, J. Sachs, and J. De Neve (eds) (2020), "*World Happiness Report 2020*". New York: Sustainable Development Solutions Network.

Hickel, J. (2021), "The Anti-colonial Politics of Degrowth", *Political Geography*, vol. 88.

Hickman, C. et al. (2021), "Climate Anxiety in Children and Young People and Their Beliefs About Government Responses to Climate Change: A Global Survey", *The Lancet Planetary Health*, vol. 5, no. 12, pp. e863–e873.

Hirvilammi, T. (2020), "The Virtuous Circle of Sustainable Welfare as a Transformative Policy Idea", *Sustainability,* vol. 12, p. 391. https://doi.org/10.3390/su12010391.

IEA. (2023), "Fossil Fuels Consumption Subsidies 2022", IEA, Paris https://www.iea.org/reports/fossil-fuels-consumption-subsidies-2022, License: CC BY 4.0.

ILO. (2016), "Guidelines for a Just Transition Towards Environmentally Sustainable Economies and Societies for All", International Labour Organization, Geneva.

IPBES. (2019), "Global Assessment Report on Biodiversity and Ecosystem Services of the Intergovernmental Science-Policy Platform on Biodiversity and Ecosystem Services", IPBES.

IPBES. (2022), "Summary for Policymakers of the Methodological Assessment of the Diverse Values and Valuation of Nature of the Intergovernmental Science-Policy Platform on Biodiversity and Ecosystem Services".

Kathiresan, K., and Rajendran, N. (2005), "Coastal Mangrove Forests Mitigated Tsunami", *Estuarine, Coastal and Shelf Science*, vol. 65, no. 3, pp. 601–606.

Klinenberg, E. (2002), *Heat Wave: A Social Autopsy of Disaster in Chicago.* Chicago: University of Chicago Press.

Laurent, E. (2022), "Air (ine)Quality in the European Union", *Current Environmental Health Reports*, vol. 9, no. 2, pp. 123–129.

Laurent, E. et al. (2022), "The Wellbeing Reflex: Facing Covid with a 21st Century Compass", WEAll Policy Brief.

Laurent, E., and Morand, S. (2021, October 11), «Bio-croissance ou biosolidarité ? La Convention sur la diversité biologique à l'heure des choix », *The Conversation*.
Marmot, M., Allen, J., Boyce, T., Goldblatt, P., and Morrison, J. (2020), "Health Equity in England: The Marmot Review 10 Years On". Institute of Health Equity. https://www.health.org.uk/publications/reports/the-marmot-review-10-years-on.
Martinez, R., Lloyd-Sherlock, P., Soliz, P. et al. (2020), "Trends in Premature Avertable Mortality from Non-communicable Diseases for 195 Countries and Territories, 1990–2017: A Population-based Study", *Lancet Global Health*, vol. 8, pp. e511–e523.
Mazzochi, T. (1993). "A Superfund for Workers", *Earth Island Journal*, vol. 9, no. 1, pp. 40–41. http://www.jstor.org/stable/43883536.
Murray, C. J., and Lopez, A. D. (eds). (1996), *The Global Burden of Disease. A Comprehensive Assessment of Mortality and Disability from Diseases, Injuries, and Risk Factors in 1990 and Projected to 2000*. Cambridge: Harvard University Press.
One Health High-Level Expert Panel (OHHLEP), Adisasmito, W.B., Almuhairi, S., Behravesh, C. B., Bilivogui, P., Bukachi, S. A. et al. (2022), "One Health: A New Definition for a Sustainable and Healthy Future", *PLoS Pathogens*, vol. 18, no. 6, p. e1010537. https://doi.org/10.1371/journal.ppat.1010537.
Ostrom, E. (1990), *Governing the Commons. The Evolution of Institutions for Collective Actions*. Cambridge: Cambridge University Press.
Ostrom, E. (2011), « Par-delà les marchés et les États: la gouvernance polycentrique des systèmes économiques complexes, conférence Nobel » (trad. Éloi Laurent), in LAURENT E. (dir.), « Économie du développement soutenable », Revue de l'OFCE, « Débats et politiques », no. 120.
———. (2010), "Beyond Markets and States : Polycentric Governance of Complex Economic Systems", *American Economic Review*, vol. 100, no. 3, pp. 641–672.
———. (2012), "Nested Externalities and Polycentric Institutions: Must We Wait for Global Solutions to Climate Change Before Taking Actions at Other Scales?", *Economic Theory*, vol. 49, pp. 353–369.
Pantell, M., Rehkopf, D., Jutte, D., Syme, S. L., Balmes, J., Adler, N. (2013), "Social Isolation: A Predictor of Mortality Comparable to Traditional Clinical Risk Factors", *American Journal of Public Health*, vol. 103, no. 11, pp. 2056–2062.
Quintanilla, M., Guzmán León, A., Josse, C. (2022), "The Amazon Against the Clock: A Regional Assessment on Where and How to protect 80% by 2025". https://amazonia80x2025.earth/.
Raworth, K. (2018), *La Théorie du Donut. L'économie de demain en 7 principes*, translated Laurent Bury. Plon, Paris.

Sandifer, P. A., Sutton-Grier, A., and Ward, B. P. (2015), "Exploring Connections Among Nature, Biodiversity, Ecosystem Services, and Human Health and Well-being: Opportunities to Enhance Health and Biodiversity Conservation", *Ecosystem Services*, vol. 12, pp. 1–15.

Snyder-Mackler, N., Burger, J. R., Gaydosh, L., Belsky, D. W., Noppert, G. A., Campos, F. A. et al. (2020), "Social Determinants of Health and Survival in Humans and Other Animals", *Science*, vol. 368, p. 843. https://doi.org/10.1126/science.aax9553.

The US Burden of Disease Collaborators. (2018), "The State of US Health, 1990–2016: Burden of Diseases, Injuries, and Risk Factors Among US States", *JAMA*, vol. 319, pp. 1444–1472.

World Health Organization. Regional Office for Europe. (1994), *Declaration on Action for Environment and Health in Europe: Second European Conference on Environment and Health: Helsinki, Finland, 20–22 June 1994*. World Health Organization. Regional Office for Europe. https://apps.who.int/iris/handle/10665/197626.

CHAPTER 4

Policy: Sustainable Pathways

From Growth Policy to Well-Being Policies

According to the International Union of Geological Sciences (IUGS), the professional organization that defines the Earth's time scale, the contemporary geological period belongs to an epoch called the Holocene (the "new whole"), which began 11,500 years ago after the last ice age. But Paul Crutzen and Eugene Stoermer (2000), wanting to emphasize the central role of man in geology and ecology, proposed to use the term "Anthropocene" to designate the current geological epoch (in Greek, *anthropos* means human and *kainos* means recent) and suggest that it begins in the late part of the eighteenth century. They chose this date because, during the last two centuries, the global effects of human activities have become clearly perceptible. Even more precisely, they chose the year 1784, when James Watt invented the steam engine. The choice of this new era is therefore technological: the successive waves of innovation and economic growth generated by the first industrial revolution unleashed the power of human domination on the planet.

But, on May 21, 2019, the members of the International Society of Stratigraphy's Anthropocene working group gave a different opinion by adopting by an overwhelming majority (88%) two proposals: the "Anthropocene" is indeed a new chrono-stratigraphic era; but this begins

É. Laurent, *Toward Social-Ecological Well-Being*, Palgrave Studies in Environmental Sustainability, https://doi.org/10.1007/978-3-031-38989-4_4

in the middle of the twentieth century[1] (i.e., later than Crutzen and Stoermer's proposal). The choice of this dating is explained, according to the specialists called upon to settle this question, by the appearance between the 1940s and 1960s of a "stratigraphic signal", carbon 14 (14C), a radioactive isotope dispersed throughout the world by the detonations of man-made nuclear weapons. This choice also coincides, as we have seen, with the inflection of many marker indicators of the Great Acceleration of the transformation of the biosphere by economic systems. But as much as the dating seems irrefutable (the middle of the twentieth century), the event chosen is insignificant in view of the three major ecological crises which have been constantly accelerating since then: climate disruption, degradation of ecosystems, and destruction of biodiversity. What is more, the choice of nuclear radioactivity as a signal of human domination refers, like the invention of the steam engine, to the power of technology, but obliterates the issue of the social dynamics that underlie it and occupy us in this book.

In 2013, Jason Moore had already challenged the idea of Anthropocene as theorized by Crutzen and Stoermer, taking humanity as an undifferentiated whole, and had pleaded for the adoption of the term "Capitalocene" (Moore 2013), "understood as a system of power, profit and re/production" bursting into the "web of Life" (this is the expression Charles Darwin used to designate the living world). But one can dispute this dating itself: the invention of capitalism is seven centuries old and the advent of industrial capitalism dates back to the beginning of the nineteenth century. However, the ecological crises really began in the 1940s and 1950s.

We can therefore choose a third way and include human domination on the planet in another historical perspective, whose name and dating differ: the advent of the "Growthocene", the era of global economic growth, which opens July 1, 1944 at the inauguration of the Bretton Woods Conference (New Hampshire, United States). If this event is foundational, it is because it consecrates gross domestic product (GDP) as a measure of the development of nations, will therefore lead to its adoption as a reference indicator by all the countries of the planet to this day and

[1] Results of binding vote by AWG. Released 21 May 2019, Working Group on the 'Anthropocene', Subcommission on Quaternary Stratigraphy, http://quaternary.stratigraphy.org/working-groups/anthropocene/.

thus induce the unprecedented environmental destruction that the curves of the Great Acceleration effectively reflect.

It was in 1934 that Simon Kuznets, a development economist, responding to an order from the United States Congress, presented the first model of what would become GDP: it was then a question of measuring the systemic economic shock generated by the 1929 crisis that sectoral economic indicators only partially and imprecisely reflect. This first GDP was improved in the midst of the war effort by a team of British economists around John Maynard Keynes, then adopted as a common standard of development at the Bretton Woods conference in July 1944. Economic policy was deeply marked by the invention of GDP in the context of the Great Depression of the 1930s in the United States, which became the supposedly effective instrument for achieving collective well-being (Laurent 2018).

GDP measures the production of goods and services exchanged on markets and monetized during a given period, by counting the flow of income, expenditure or added value. "Economic growth" refers to the increase in the level of GDP at constant prices (we subtract inflation from nominal GDP to obtain growth in volume). The "standard of living" or "level of development" can be assessed by dividing GDP by the population; "economic development" is measured by the increase in GDP per capita from which population growth is deducted.

Therefore, by construction, GDP and growth reflect only a small part of the determinants of human well-being (by measuring health and education only very poorly, for example), and in no way its resilience (its ability to withstand shocks) or its sustainability (its future evolution under ecological constraints). Kuznets himself had had the intuition of this as early as 1934 when he warned members of Congress: "Those who ask for more growth should specify their thinking: more growth of what and for what?" (Laurent 2018).

GDP and its growth are questionable from many angles but their greatest flaw is that they no longer fulfill their purpose as an indicator, which consists on the one hand of providing information on the state of the world and on the other hand of indicating a desirable direction to follow (the index which shows the way). It is thus striking to note the yawning and growing gap between GDP growth and human development. GDP and its growth superficially reflect the "wealth of nations" but are not its root cause: the central indicator, in the eyes of Adam Smith, was not GDP but labor productivity, from which economic growth flows but whose increase draws a public policy horizon distinct from that of the increase in GDP. Public health and education policies thus appear, in the light of Smith's analysis, as priorities for increasing labor productivity, whereas they are marginalized in current economic systems that are obsessed with GDP growth and fed on the expansion of finance, digital industries and fossil fuels. We can go further and reverse the hypothesis usually advanced about the remarkable increase in the standard of living in the West of the twentieth century. It is the even more remarkable increase in health conditions and levels of education that has supported the increase in labor productivity and, ultimately, that of GDP per capita.[2] GDP thus appears retrospectively and not only prospectively as a superficial indicator of human development (understood as the equally weighted increase in per-capita income, health and education) with regard to these deep determinants.

While the correlation between GDP growth and human development is much lower than is often argued in the public discourse (Laurent 2021), the correlation between GDP growth and environmental degradation is clear, which explains why this criticism of GDP and growth as intangible economic horizons are flourishing at a time when ecological crises are accelerating (see Chapter 1) and why post-growth studies have greatly developed in recent years.

In an attempt to organize this burgeoning academic field, one may wish to distinguish three schools of exit from growth that have emerged or have developed strongly over the past five years: degrowth, the "Donut Economy" and the well-being economy.

Degrowth has its origins in the work of the Romanian-born economist Nicholas Georgescu-Roegen in the 1970s, but it has recently undergone a theoretical and empirical overhaul. A contemporary definition of this stream has been given by one of its most active French researchers, Timothée Parrique (2019), in the following form: "The planned and democratic reduction of production and consumption in rich countries, to reduce environmental pressures and inequalities, while improving the

[2] Data compiled by Prados de la Escosura (2015) show that improvements in health and education account for 85% of the increase in the human development index over the past 140 years, both for OECD countries and European countries than for the rest of the world. On the other hand, human development is closely linked to the expansion of the welfare state, the most humanly developed countries in the world being those which were able to invest very early in their welfare state (Laurent 2021).

quality of life". We could speak, in the light of the work of researchers of this stream, of a dialectic of degrowth: just as it seems impossible to them to decouple economic growth from the environmental damage it causes, so it seems possible to them to satisfy universal decent standards of living while reducing global energy consumption (which has doubled over the past 40 years), on the condition of a gigantic redistribution of resources between and within countries.

The Donut Economy stemmed from the work of British economist Kate Raworth, who in a 2012 Oxfam working paper proposed adding a "social floor" (formed of eleven societal variables such as "food security, water and sanitation or health care" to the "ecological ceiling" made up of nine "planetary boundaries". "Between a social floor that protects against critical human deprivation, and an environmental ceiling that avoids exceeding critical natural thresholds, lies a safe and just space for humanity—shaped like a doughnut. It is the space where human well-being and planetary well-being are assured and where their interdependence is respected … a space within which humanity can thrive" writes Raworth (2012).

Finally, the Well-being Economy Alliance (WEAll) is an international network, created in 2018 and made up of associations, researchers and political leaders, that promotes the "well-being economy", disseminating available knowledge on alternative economic models centered on human well-being with the explicit aim of going beyond GDP and economic growth (Abrar 2021) and working directly with national or territorial public authorities to translate them into well-being policies (Table 4.1).

A quick bibliometric review, based on occurrences of these different schools in English-language works from 2000 to 2019 using Books Ngram Viewer, shows that the revival of work on post-growth coincides with the aftermath of the Great Recession (work of which the Stiglitz Commission of 2008–2009 was the symbol) but also that the degrowth trend is by far the most visible in academic literature today. But these streams are not watertight: their authors read, collaborate and criticize each other constructively (the degrowth authors being the most collaborative). There is therefore an overall movement driven by these fundamentally converging currents.

Thus, while the authors of degrowth have empirically demonstrated the incompatibility between the pursuit of GDP growth and the preservation of the biosphere and insist on the possibility of satisfying essential human needs with reduced consumption of energy and natural resources

Table 4.1 Three post-growth streams

	Founding date	*Theoretical advance*	*Empirical advance*	*Institutional progress*	*Main contributions*	*Main representatives*
Degrowth	1972/2008	+ +	+ + +	+	Decoupling/essential energy needs	J. Steinberger/G. Kallis/J. Hickel
Donut economy	2012	+	+ + +	+ +	Social floor/ecological ceiling	K. Raworth/D. O'Neill/G. Thiry-P. Roman
Well-being economy alliance	2018	+ +	+	+ + +	Well-being Budgets	K. Trebbeck/L. Fioramonti/I. Kubiszewski

Note the + sign qualifies the intensity of the progress according to the author's assessment
Source Own elaboration

provided there is massive redistribution, these same essential needs have been specified theoretically and articulated empirically with the planetary limits by the Donut Economy. At the institutional level, there is also complementarity between the work on integrating well-being indicators at the local level undertaken by the Donut Economy and this same work undertaken at the national level by the Alliance of Well-Being Economies. Finally, of course, all these currents agree on the social and environmental limits of GDP growth and the need to develop alternative visions and operational indicators.

The concept of these converging streams and the associated effort has been taken up by a number of key bodies of the international community, inducing a second rise in power of an institutional nature, which began with the adoption of the seventeen Sustainable Development Goals (SDGs) by the United Nations in September 2015. We have already noted that the first part of the IPCC's AR6 report published in August 2021 aims for a world in which "the emphasis on economic growth shifts in favor of human well-being" (IPCC 2021), but this overtaking of economic growth is also enshrined in the 2021 joint report of the IPCC and the IPBES, which recommends "moving away from a conception of economic progress where only GDP growth prevails" (Pörtner et al. 2021) to preserve biodiversity and ecosystems. The European Environment Agency (EEA 2021) has published a note recommending going beyond economic growth to achieve the objectives of the European Green Deal, while the OECD continues its work of promoting well-being indicators with the launch in 2021 of a new research center dedicated to these themes (Centre on Well-being, Inclusion, Sustainability and Equal Opportunity, or WISE).

Finally, these two upsurges in academic and institutional spheres have been accompanied by a political translation: more and more countries and localities are developing both indicators and well-being policies (Fig. 4.1).

This is particularly the case recently in Iceland, Scotland, Finland and New Zealand, members of the Wego program launched by the Alliance of Well-Being Economies (Abrar 2021), and more recently Germany. At the local level, the cities of Amsterdam and Brussels are at the forefront of well-being policies.

How to solidify and institutionalize the vital shift away from economic growth toward human well-being in analysis and policy alike? First, by defining well-being and then by focusing not just on well-being indicators but on actual well-being policies.

If cooperation is the means to human prosperity (see Chapter 1), the

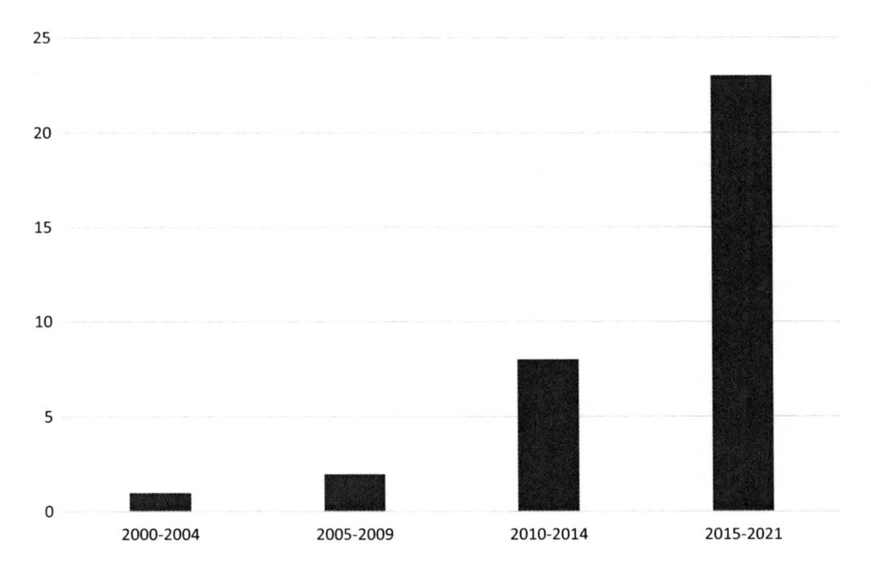

Fig. 4.1 Number of countries that have launched well-being initiatives, 2000–2021 (*Source* OECD, own elaboration)

goal of human interactions can be broadly defined as well-being. The root of this term is ancient and its ramifications are numerous. The philosophers of ancient Greece and, more specifically, the proponents of the "eudaemonist" stream (from the Greek *eudaimonia,* which means happiness) wondered more than two millennia ago about the link between individual happiness, collective prosperity and moral virtue. Both Thomas Jefferson and Jeremy Bentham made well-being the foundation of political liberalism in modern times. Finally, the welfare state became the heart of public power in the twentieth century. Well-being can be defined simply as a personal or social state of feeling good (subjective dimension) and doing well (objective dimension). It can be measured according to different dimensions (health, education, environmental quality, etc.) using different indicators (life expectancy, level of education, air quality, etc.). The discipline of economics has developed different streams of "welfare economics" (first centered on the aggregation of individual utilities, then on the question of the coherence of collective preferences, up to contemporary theories of justice with Rawls and Sen).

Recently, well-being has been connected to sufficiency, an ancient idea of good living and moderation. Economic analysis was actually conceived by Aristotle two and a half millennia ago as a discipline of sufficiency, seeking to satisfy essential human needs in a constrained environment by ensuring the correspondence between reasoned needs and limited resources. But this first sufficiency, which is close to the notion of personal frugality, unfolds in the space of the home which is by nature unequal: the members of the family are placed in a hierarchical relationship and must not become equals. There is, therefore, no reason for the satisfaction of basic needs, which proceeds from a principle of necessity, to result in a just situation. It is in the space of the city (*polis*) that the necessary may or may not be judged as sufficient.

How to understand basic human needs? in his *Letter to Menoeceus*, Epicurus classifies desires in three categories. (1) "natural and necessary desires" (for the well-being of the body (*aochlèsia*): to protect the body against the aggressions of the cold and bad weather, fire, clothing and shelter will be needed; for happiness (*eudaimonia*); for life itself; "vital needs" such as hunger, thirst, inclination to rest (sleep). (2) "natural and unnecessary desires": sexual desire and aesthetic desires; "vain desires (*kenai*)" are those that go beyond the limits inherent in nature such as the thirst for possession, the thirst for power and the thirst for honors.

The second age of sufficiency, taking note of the dazzling acceleration of economic development after the Second World War, intended to slow down the overconsumption of the natural resources it engendered, starting with the supply of energy. French NGO Négawatt introduced this concept of sufficiency moderation in the early 2000s to distinguish it from the logic of energy efficiency. As much as energy efficiency aims to reduce the quantity of energy (and/or carbon) per unit of production, energy sufficiency (*sobriété* in French) aims to reduce the volume of energy consumed and therefore to guarantee that the policies implemented are reflected effectively through energy savings by avoiding a "rebound effect" in consumption. Négawatt's visionary work has made it possible in particular to distinguish between different sufficiency policies: structural (e.g., bringing places of work and residence closer together), dimensional (e.g., reducing the size of cars), usage (e.g., taking public transport) and collaborative (e.g., promoting car sharing).

Building on the work of Négawatt, sufficiency was for the first time clearly defined in the high-level climate reports when the IPCC in its 2022 Summary for Policymakers mentioned "sufficiency policies" as "a set of measures and daily practices that avoid demand for energy, materials, land and water while delivering human well-being for all within planetary boundaries". The report also provided a rather more precise formulation on the concept in Chapter 9 on Buildings, where it is stated:

> Sufficiency addresses the issue of a fair consumption of space and resources … The remaining carbon budget, and its normative target for distributional equity, is the upper limit of sufficiency, while requirements for a decent living standard define the minimum level of sufficiency… Decent living standards are a set of essential material preconditions for human well-being which includes shelter, nutrition, basic amenities, health care, transportation, information, education, and public space.

Two elements are conflated in the IPCC definition: the conceptual insight of sufficiency ("human wellbeing for all within planetary boundaries") and policies aimed at lowering demand ("measures and daily practices that avoid the demand for energy, materials, land and water"). The IPCC is also breaking down sufficiency policies that are targeting demand along a triptych: avoid, shift, improve.[3]

[3] Hickel et al. (2021).

Because sufficiency is an equilibrium or optimal state between poverty and excess, it resonates well with the language of the post-growth agenda. Echoing the work of Epicurus, British scholar Ian Gough (2021) offers a contemporary vision of human aspirations, based on needs and not preferences, and thus manages to define essential human needs based on three principles:

1. Human needs are universal. All people, everywhere, present and future, have certain basic needs, in particular health, autonomy and participation. These basic needs in turn induce a range of intermediate (second-order) needs, both material (such as nutritious food, protective housing, health care, educational pathways) and immaterial (such as quality relationships). Such needs have five characteristics: they are objective, plural, non-substitutable, satiable (susceptible to being satisfied) and transgenerational.
2. These universal needs differ from the processes aimed at satisfying them, which are specific to different human groups and variable in time and space (feeding is a universal need, but there is an infinite range of practical means of doing so, means attached to historical, social, cultural contexts, etc.). The "necessities" then designate all the goods and services considered as an acceptable minimum to satisfy human needs in a given society, which distinguishes them from luxury.
3. The identification of needs requires a specific collective process distinct from the processes of revealing preferences on the markets, such as citizen forums.

Recent empirical work has attempted to clarify the quantification of these universal essential needs with regard to environmental constraints, and in particular the need to reduce energy consumption to stem the climate crisis. A first study (Kikstra et al. 2021) attempts to characterize what a "decent living standard" would be on a global scale (including health, housing, food, socialization and mobility) and manages to show that more people in the world are deprived of it than people considered to be financially poor (which is in line with the work of Amartya Sen on the need to consider poverty more broadly than through the monetary prism alone). More fundamentally, the study shows that the cumulative energy needs corresponding to the construction of new infrastructures necessary to ensure a decent standard of living for the entire world population by 2040 represents less than three-quarters of the world's annual energy demand today. The authors conclude that the eradication of poverty does not constitute, in itself, a threat to the mitigation of climate change on a global scale and that it is essential to distinguish between comfort energy and energy luxury for a decent life when developing just energy strategies under climate constraints.

A second study (Vogel 2021) is interested in the question of the means mobilized to satisfy these essential human needs in energy under climatic constraints and identifies in particular three beneficial socio-economic factors in this perspective: the quality of the welfare service, income equality and democracy, which are associated with lower and better satisfied energy needs.

There is therefore a theoretical base and empirical underpinnings to well-being, understood in the context of sufficiency.

From this definition of well-being, which should be adapted to fit national and local preferences, well-being policies must be developed, as has been the case in New Zealand since 2019 (Box 4.1).

Box 4.1: New Zealand's Well-Being Budgets

On May 30, 2019, New Zealand decided to place its public finances under the aegis of well-being defined as follows: "Well-being means that people are able to lead a fulfilling life with a purpose, a balance and meaning for them. Empowering more New Zealanders to enjoy that well-being requires tackling the long-term challenges we face as a country, such as the mental health crisis, child poverty and violence. domestic."

The design and implementation of this well-being budget resulted from three fundamental principles:

1. Moving away from the silos of public agencies and work across government to assess, develop and implement policies that improve well-being. An example of this change is that the Treasury now requires collaboration between ministries when submitting budget proposals. This has led, for example, to ten agencies coming together to jointly try to help combat domestic and sexual violence.
2. Focusing on results that meet the needs of present generations while thinking about their long-term impacts for future generations. Based on this, the government has identified five priority well-being

areas for the 2019 budget: improving mental health, reducing child poverty, tackling inequalities faced by Indigenous Maori and Pacific Islanders, thriving in the digital age, making the transition to a sustainable low-carbon economy.

3. Monitoring the progress of these policies using appropriate indicators, in particular individual and community health. This involves revamping budget documents to clarify how any policy or initiative, including the government balance sheet and the management of public assets, has contributed to improved well-being.

The 2019 well-being budget was institutionalized by a modification of the law on public finances in 2020 and extended in 2021 and 2022 in particular to better take into account the question of environmental sustainability.

Source https://www.treasury.govt.nz/publications/wellbeing-budget/wellbeing-budget-2020-html

Actually, a distinctive feature of well-being economy studies has been their policy-oriented nature, with WeALL (Janoo et al. 2021) putting out a "Wellbeing Economy Policy Design Guide—How to design economic policies that put the wellbeing of people and the planet first" with five key steps toward a well-being economy:

1. Develop a well-being vision, framework and measurements.
2. Design a strategy to foster the areas of economic life most important for our well-being.
3. Assess and co-create Wellbeing Economy policies to build a coherent and innovative policy mix.
4. Successfully implement Wellbeing Economy policies by empowering local stakeholders and communities.
5. Evaluate policy impacts on well-being for learning, adaptation and success.

In the same operational vein, the WHO is currently developing a new policy agenda around the well-being economy, in connection to health.[4]

[4] On March 1–2, 2023, WHO/Europe held a "Health in the Well-Being Economy" Regional Forum, with the definition "a well-being economy places people and planet at

With Covid, we have entered the century of ecological shocks. In this new century, well-being can serve both as a compass and a shield: pointing to meaningful collective horizons while building substantial collective resistance. In this respect, "well-being indicators" are not amusing gadgets that cannot make any significant changes to core economic policies: if they are embedded in policies and institutions, they can help foster a well-being culture that can make a huge difference in terms of human well-being in time of crisis, as they did in many countries around the world in 2020–2021.

From this perspective, the age of "indicators" is behind us: we now need to work on well-being policies, i.e., operationalizing new visions of the economy and mainstreaming these visions into policies.

From Cost–Benefit to Co-Benefits

At first sight, in the conventional conception of the economy, everything seems relatively simple for the public decision maker: the choice of an environmental policy (aiming for example at reducing GHG emissions) will depend on the comparison between its social cost and its social benefit here and now. If the benefit exceeds the cost, the policy will be implemented by the political authority in the name of the general interest: this is what is known as "cost–benefit analysis".

This decision-making aid technique, widely used by public authorities around the world in environmental matters, has its consistency (finding a common unit of account to compare dimensions that are not spontaneously shared, such as the use of a side and the health of the other) and may have its uses. Thus, when the Obama administration tried to commit the United States to a low-carbon energy transition (with the 2015 Clean Power Plan), it was able to rely on a cost–benefit analysis by the Environmental Protection Agency (EPA) showing that this strategy, in particular due to the health gains associated with phasing out coal, was largely beneficial (a study that the Trump administration later removed from the EPA website).

But cost–benefit analysis (CBA) poses two major problems. It leads to defining on the one hand the end to be achieved (the maximization of monetary gain) and on the other, at the same time, the means to be

the centre of creating healthy, fairer and more prosperous societies" and "securing resilient health systems".

deployed to do so. We can therefore prefer a "cost-effectiveness analysis" which is limited to defining the most efficient way from the point of view of the use of available resources to achieve a desired objective (for example, the preservation of a certain number of species of birds in a territory). This difference gave rise in 2021 to a skirmish between two groups of economists on the legitimacy of the "social cost of carbon" calculated from the cost–benefit analysis (on the one hand, Stern and Stiglitz [2021]; on the other, Aldy et al. [2021]). One can also rely on multi-criteria analyses, which do not confine environmental policies to the sole objective of economic efficiency but also consider dimensions such as security, sustainability and, of course, justice.

The other major weakness of CBA is that since it aggregates monetary gains and losses garnered or borne by different individuals and groups, it omits issues of distribution and therefore of justice induced by public decisions and can therefore logically lead to unequal and/or counterproductive recommendations.

The most famous example of these errors of analysis is the recommendation made by the chief economist of the World Bank, Larry Summers, in the early 1990s to dump toxic waste in low-income countries: given the lower wages of workers in developing economies, the monetary cost of harming their health and possibly ending their existence prematurely by polluting their environment indeed appeared to be lower than the monetary cost of maintaining polluting industries in developed countries (which would risk poisoning better-paid workers). Here we find the ambition of an amoral economic reasoning, which comes up against the gaping methodological limits of the calculation of the value of a human life.

But, when justice, rather than narrowly defined "efficiency", is seen as a goal, the cost–benefit "logic" breaks down: human beings have an equal right to be healthy and simply alive, regardless of their nationality, their payslip and, more broadly, their living conditions. On the other hand, when we consider the dynamic global perspective (i.e., over time), this "logic" becomes counterproductive: allowing the most polluting industries to be outsourced from rich countries to poor countries means that the pollution in question is never reduced, but simply transferred, leading for example, in the case of CO_2-emitting industries, to an increase in global GHG emissions, which end up harming the health of workers in developed countries with the climate disruption they cause. And yet, even today, by virtue of the same "economic logic", the OECD continues to base its public policy recommendations (for example on the fight against air pollution) on monetary valuations of human lives which result in a gap of 1 to 3.5 between a Chinese life and an American life or a simple doubling between a Brazilian existence and a French existence.

But the flaws in cost–benefit analysis run even deeper. The horizon of environmental policies is a long one (the end of the century in the case of climate policies). The comparison between the costs and benefits of environmental policies must therefore relate to several successive generations in order to have any real meaning.

This projection in time presupposes a calculation of actualization, that is to say a translation into the language of the present of future events and their consequences. The discount calculation, commonly used to assess the profitability of private investments, must also be used to analyze and quantify the consequences of climate change, but also the loss of biodiversity or the depletion of ecosystems from a collective perspective. We then speak of "social discounting" and, more precisely, of the "social discount rate" (Box 4.2).

Box 4.2: The Social Discount Rate and Its Variables
According to the formula of Frank Ramsey (1928): $R = p + e*g$:

- R is the discount or social discount rate: the higher it is, the less the costs imposed by our actions on future generations seem significant to us; the smaller it is, the more the well-being of future generations is on an equal footing with ours.
- p (pure rate of time preference) represents the impatience of current generations to consume the resources at their disposal, knowing that 1 euro of immediate consumption is preferred to 1 euro of future consumption; $p = 0$ would mean an indifference to the passage of time, to the date on which one consumes.
- e (elasticity of the marginal utility of consumption) represents both aversion to inequality between generations and, under certain assumptions, aversion to fluctuations in consumption (therefore for risk); if $e = 1$, inequality aversion is very low (we do not care how consumption growth is distributed over time between generations).
- Finally, g (per-capita consumption growth rate) represents the increase in consumption from which future generations will benefit; if $g > 0$, we expect future generations to be richer than us.

Let us start from the present value of a private investment, for example the purchase of a taxi license in a large metropolis. The price of this license (supply side) and the value of this investment (demand side) will depend on the anticipated income streams that the exercise of the taxi business allows. Thus, on average, over the past ten years, the price of a taxi license in France has been halved under the pressure of competition from VTCs and other platform services which automatically reduces the anticipated flow of trips for taxi drivers. We can talk about the net present value of the investment consisting in buying a taxi license (around €150,000 in a city like Paris). But how to actualize this socially? What is the discounted value of a collective/social investment (on this subject we speak of "shadow value")?

If the interest rate is the price of time, then the social discount rate can be understood as the value placed on the well-being of future generations relative to the value placed on the well-being of present generations. However, because social discounting is applied over a long period, it results in very significant valuation differences: a social discount rate of 2% applied to €1 million of economic cost amounts to €552,000 after 30 years and to €138,000 after a century; and a rate twice as high (4%) results in a reduction of the same amount to €308,000 after 30 years and barely €20,000 after a century, i.e., a 98% reduction on the original cost. Discounting can therefore be misleading: costs considered at sufficiently distant dates with a positive, even very low, social discount rate minimizes the costs of our decisions for future generations and maximizes our benefits. Discounting clouds and possibly obliterates our responsibility.

Because we choose social discount rates that are too low for dubious ethical reasons, we end up minimizing the efforts required of us in the short term. In fact, the academic literature retains positive values, sometimes high ones, for the social discount rate, giving the illusion that the costs of environmental crises are bearable on condition that the resulting inequalities are minimized and that the emphasis is on the future growth of monetary income.

Cost–benefit analysis essentially leads to opposing the different dimensions of human well-being and the well-being of different generations by claiming to reconcile them in the language of material interest. On the contrary, "co-benefits" analysis makes it possible to reason in a continuous and integrated way.

To understand how cost–benefit analysis can lead to policy aberrations, let's consider the price of a life in France during Covid. A French life would be worth €3 million, is the conclusion of a 2016 report taking up the conclusions of a 2012 report by the OECD, which aggregated dozens of disparate studies to arrive at this improbable figure.

In 2020, France experienced 65,000 deaths due to Covid (30,000 during the first wave in spring, 35,000 during the second wave in autumn). At €3 million per life, this represents €195 billion in losses. While the lockdowns limited the number of infections and deaths, at the same time they resulted in a drop in gross domestic product (GDP) of around 8%, or €210 billion. It is therefore an almost perfectly balanced financial operation from an accounting point of view. But, between aid to ailing businesses and recovery plans, the French state had to pay an additional €100 billion. Health policy in France has therefore not been optimal: life has been "overpaid", very precisely by 380,300 units.

Suppose now that this debate had indeed taken place during the first wave of Covid and that the government, worried about the magnitude of the potential loss of production and duly informed by ingenious economists, had opted for an easing of the restrictions. Mechanically, there would have been more deaths (that would be the goal, if we dare to say so), but also an explosion of infections and incidence rates, so that it would have been necessary, after a period of denial, to impose stricter lockdowns for even longer to contain the pandemic, which would have reduced production even more. We would thus have, under the pretext of giving priority to "economy" over "health", lost on both counts. This is what happened in a neighboring country governed by radical neoliberalism: the United Kingdom.

Here we touch on the economic and social absurdity of cost–benefit analyses, the principle of which is to oppose the respective monetary values of the different dimensions of human well-being. If the Covid crisis teaches us anything, it is that these economic trade-offs are in fact ethical renunciations which are proving counterproductive: it is now empirically established that governments which first sought to preserve health have also imposed fewer freedom restrictions on their populations while preserving income and employment levels. Finland and New Zealand (much less hard hit economically than Sweden or the United Kingdom, for example) saw their life expectancy increase in 2020, while it fell sharply in OECD countries, and especially in France. Conversely, the disaster was as much economic and social as health and political in the United States, Brazil or India.

In fact, the early narratives concerning the public health policy response to Covid often conveyed the idea that optimality-seeking policymakers needed to account for trade-offs between different dimensions of human well-being. In particular, a public health response that prioritized safeguarding people from the virus would have adverse impacts on other aspects of well-being. This led to policy advice proposing that public health policy should accept a higher death rate as part of a trade-off with other domains of well-being (a recent study, Laurent et al. 2022, has been able to identify three variants of this trade-off narrative, see Box 4.3).

Box 4.3: The Trade-off Narratives of Covid

First narrative: health vs "freedom"

In the early weeks of its emergence, the fast-developing Covid pandemic triggered a first narrative based on an alleged trade-off between civil liberties and social control of infections. Based on the very first lockdown experience imposed on millions by the authoritarian Chinese regime in the megacity of Wuhan on January 23, 2020, democratic nations were invited to follow suit and acknowledge the fact that to mitigate Covid, freedom had to be mutilated.

The first, essential, problem with this narrative lies in its definition of freedom, which cannot be equated to the right to impose the negative consequences of its own behaviors on others, but actually includes responsibility toward the community, a principle at the heart of public health policies. In this sense, a restriction on freedoms in the face of a global pandemic can actually be liberating, while a *laissez-faire* approach can be constraining.

These theoretical considerations lead to the second, empirical, problem of this narrative: countries that prioritized health ended up imposing lesser restrictions on freedoms as measured by the Oxford Stringency Index. Oliu-Barton et al. (2021) have shown that, on the contrary, countries that neglected health indicators and imposed lighter restrictions on economic activity (remote work, curfews, etc.) had to ultimately impose greater restrictions on civil liberties and political rights (freedom of assembly, mobility, etc.).

Second narrative: health vs "the economy"

The second narrative that emerged early in the pandemic claimed that countries were confronted with a blunt but unavoidable choice between "saving lives" and "saving the economy" (or, as the *Financial Times* put it on a graph plotting the loss of lives and the loss of production measured by GDP, between preserving "lives" or "livelihoods").

Two years down the road, the initial claim that countries could either save their economies or save lives finds itself invalidated by the most comprehensive studies. There has been no trade-off between the economy and health since March 2020: either countries have preserved both, or they have hurt both. A recent Molinari Institute report shows that France is the very counter-model of the health–economy double penalty (Philippe and Marques 2021). Countries that prioritized health ended up winning on all fronts (New Zealand, Finland, Iceland, etc.), while countries that prioritized economic activity lost on all fronts (US, UK, Italy, France, etc.).

Third narrative: mental health vs physical health

The third narrative, that took hold even later with lockdown fatigue, seems more convincing than the two previous ones. It argues that saving lives can come at the price of harming minds. Emergency lockdown policies implemented to avoid hundreds of thousands of deaths are policies of de-socialization which came at an exorbitant cost for well-being, starting with mental health and happiness, which depend on social life and social bonds. In theory, thus, lockdown policies pit physical health against psychological health: repeated and prolonged restrictions on social cooperation and social bonding strongly affect psychological balance and personal happiness.

There is in fact ample evidence that mental health has seriously suffered during Covid because of lockdowns (OECD 2021), which makes it even more important to limit lockdowns in time. This is precisely why, here again, trade-offs appear misleading. As we have seen, countries that have prioritized physical health ended up imposing stricter but shorter restrictions on social cooperation so that mental health, indeed critically dependent on social ties, was better preserved. For instance, France ended up imposing 60% more restrictions than New Zealand during March 2020 and March 2021.

Source adapted from Laurent, E. et al. (2022).

Yet, in the analysis they offer of their own data for the twelve months from March 2020 to March 2021, the Oxford researchers who developed "The Oxford Covid-19 Government Response Tracker (OxCGRT)" (Hale 2021) designate six countries that proved unable to design an effective and coherent strategy against COVID-19 and therefore were tossed from one wave to the other, oscillating between periods of brutal restrictions and *laissez-faire*: the United States, the United Kingdom, South

Africa, Iran, Brazil and France. Conversely, they highlight the health and economic successes of countries markedly less well endowed in health care capacities, such as Mongolia, Thailand and Senegal.

Even more, case studies of success stories of mitigating the Covid impact reveal that some nations that have developed over recent years a real commitment to integrating human well-being in their public policies (for instance New Zealand, Finland and Bhutan), i.e., a well-being culture, have had a "well-being reflex" that allowed for a well-being-based response to Covid, resulting not only in the preservation of human life and the various dimensions of human health but also in important social, economic and political co-benefits (Laurent et al. 2022).

Hence, we should move beyond the cost–benefit approach which continues to dominate our collective actions and decision making. This approach assigns a monetary value to every aspect of life and evaluates our actions and investments in terms of their relative monetary cost vs their relative monetary benefit.

Instead, a co-beneficial approach recognizes the intrinsic value of the health of our people and planet and their role as the foundation for any economic activity. In this latter perspective, mitigating climate change is not only vital for our collective health and well-being, but it also brings about considerable social savings resulting from improved health, as well as economic gains.

Co-benefits were defined fifteen years ago by the Health and Climate Change Commission set up under the aegis of the medical journal *The Lancet* as the collateral benefits linked to the reduction of GHG emissions, such as improved air quality, technological innovation or job creation. For instance, when all co-benefits are taken into account, the switch to renewable energies would lead to savings of around fifteen times the cost of their deployment. The study by Mark Jacobson and his team of colleagues and students (Jacobson et al. 2022) covers 143 countries representing 99.7% of global CO_2 emissions. It is based on a detailed and quantified scenario of transition to 100% solar-wind-hydraulic (WWS) energy by 2050 at the latest with a share rising to at least 80% by 2030. In this scenario (which minimizes outages and blackouts), energy needs are reduced by 57%, energy costs reduced by \$17.7 to \$6.8 billion per year (61% reduction) and social costs (in particular sanitation) from \$76.1 to \$6.8 billion dollars per year (91% reduction). The 100% WWS scenario also creates 29 million more full-time jobs than are lost (particularly in fossil fuels and the nuclear industry) and requires only around 0.5% of the ground surface to be deployed.[5]

Among these co-benefits, gains in human health are preponderant but they do not actually need to be monetized to be tangible.

There are three main steps to assess health gains (or losses) associated with an improvement (or a degradation) of environmental conditions: the evaluation of exposure and risk; the economic quantification of this evaluation; and finally the monetization of this economic quantification.

The first step consists in establishing the exposure–risk functions. The exposure–risk functions are derived from epidemiological studies and make it possible to associate an exposure indicator (for example a too meaty diet) with a health risk (premature death or incidence of chronic diseases). The detailed knowledge of this association makes it possible to evaluate the number of cases attributable by risk (for example the number of hospital admissions due to air pollution in France).

The second step is economic quantification. The quantitative assessment that can be drawn is either negative or positive: the economic cost of air pollution can be calculated or, conversely, the economic gains linked to the reduction of air pollution. Finally, one can choose (or not) to monetize gains or losses. If so, monetization of human life should be used, for instance in the form of the value of a statistical life (VSL),[6] with all the limitations already noted.

But this calculation does not have to be done in monetary terms. It can seek to assess the losses (or conversely the gains) in well-being for the population: for example, the reduction in mortality linked to lung cancer, expressed either in the number of premature deaths avoided, or the life expectancy gained, or in years of life gained (in the sense of the DALY, see above).

An alternative method would be to measure not the monetary value of human lives, but the loss in life expectancy due to poor environmental

[5] These robust 100% renewable scenarios, based on energy security, economic efficiency and ecological sustainability criteria, are considered the most desirable by the IPCC (in particular because of the risks that the climate crisis and the structural droughts it generates pose to nuclear reactors, which constantly need to be cooled).

[6] We first establish the value of a year of life (VOLY), then, by multiplying by the life expectancy, we determine the "value of a statistical life" (VSL), 3 million euros corresponding to the value of a lifetime.

conditions and the social cost of treating pathologies related to environmental conditions (such as chronic respiratory diseases or, worse, cancers resulting from exposure to urban pollution).

Pioneer work in this respect is the Air Quality Life Index (AQLI) developed at the University of Chicago, which does not use life monetization but rather measures the gains in life expectancy from enhanced environmental policy (according to the AQLI, exposure to an additional 10 μg/m3 of PM10 reduces life expectancy by 0.64 years and each additional 10 μg/m3 of PM2.5 exposure reduces life expectancy by 0.98 years). More broadly, curbing air pollution, which could save 500,000 lives per year in the EU, has the immediate effect of reducing social spending here and now and in the face of future ecological shocks, such as the Covid crisis. The same applies to noise and its immediate effects on cardiovascular pathologies, or food quality and its immediate effects on physiological and psychological health (obesity and diabetes also play a key role in health vulnerability in Europe).

If we consider only the first two stages, existing studies known as HIAs (health impact assessments) distinguish:

- Tangible gains: direct gains (medical benefits, savings in medical expenses) and indirect (productivity gains linked to the reduction in sick leave and absenteeism).
- Intangible gains (well-being, life expectancy, gain in quality of life, psychological gains linked to the reduction of suffering for oneself and others).

The notion of social savings conveys the power of co-benefits (Fig. 4.2). For France, studies focusing on the cost of air pollution converge to evaluate the cost for the French healthcare system of the treatment of pathologies associated with air pollution as between €1 and 2 billion per year, which represents between 15 and 30% of the deficit of the health branch of social security. A recent study for the city of Grenoble consists in proposing to pilot the city with health-environment indicators in order to deduce public policies aimed at co-benefits and environmental justice.[7]

[7] Morelli et al. (2019).

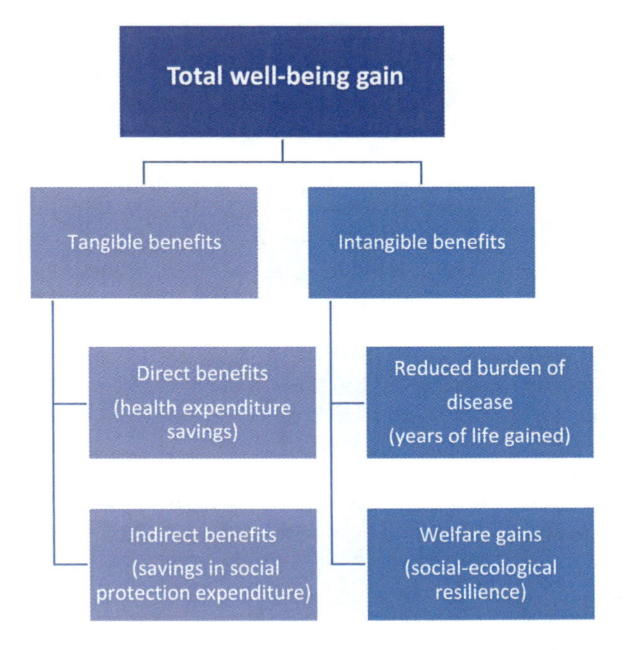

Fig. 4.2 Co-benefits and social savings (*Source* Own elaboration)

Moreover, cutting fossil-fuel subsidies and implementing progressive social-ecological taxes could be used to finance de-carbonization investments, leading to improvements in human health, savings in social spending and additional resources that could be allocated to social-ecological transition, among other things (Fig. 4.2 and Box 4.4).

A co-beneficial approach can be systematically used to improve social and ecological synergies within six key areas, in which co-benefits are both substantial and within reach (Box 4.4).

Box 4.4: Five Co-benefits Policies

1. Prioritize prevention and mitigation within and outside the health-care system—through improving the health literacy of citizens and governing bodies (who create or uphold hostile unhealthy environments)—positively affecting health, environment and the economy

by reducing the number of ill people, environmental damage and public spending.

2. Rethink food production and consumption by promoting agro-forestry practices and healthier dietary patterns associated with both lower environmental impacts and lower costs for national healthcare systems.
3. Shift towards strategies of global and national energy transition, linking health, employment, sustainability and safety co-benefits, as they provide compelling and robust evidence of short- and long-term gains
4. Invest in social cooperation as the main source of human prosperity and the key to shared human and planetary well-being, with many co-benefits for both people and planet (starting with reducing social inequality and mitigating social isolation).
5. Rethink education for the short- and long-term health of communities, by focusing on the attainment of critical thinking (rather than on the accumulation of knowledge) to be achieved via experiential learning, in connection with and learning from nature.

Source Adapted from Laurent et al. (2021)

Finally, these co-benefits are dynamic. Let us return to the issue of climate change and suppose that, out of consideration for future generations and on the basis of the best available science (i.e., the latest IPCC AR6 reports [IPCC, 2021; 2022]), we choose, finally, to halve our CO_2 emissions by 2030 by sharply reducing our consumption of fossil fuels, in other words by engaging in a real energy transition with a climate vocation. One of the consequences would be that local pollution, such as fine particles generated by automobile traffic, would also be reduced, which would immediately and sustainably improve our health and that of our children. This overlapping of the interests of generations is not a fantasy: it is exactly the path on which China has embarked. Ecological solidarity between generations, at least implicit, has been made possible there by two strategic choices: the exit from hyper-growth (drastic reduction in GDP growth) and the reduction in coal consumption. Air pollution has fallen sharply in recent years at the same time as Chinese and therefore global emissions have fallen (China being by far the largest emitter of GHGs). In this sense, future generations improve the lot of present generations, who have chosen to care about future generations.

In short, the economy of the twenty-first century is an economy of dynamic co-benefits along a social-ecological chain that links biodiversity to ecosystems, ecosystems to human health, human health to social ties, and social ties to social cooperation, of which economic activity is only one of the facets.

From Digital Acceleration to Social Continuity

California is home to two unique types of ecosystems: natural ecosystems and digital ecosystems. The former have been in structural crisis since the early 2010s: drought, fires, floods, air pollution, hurricanes, etc. The latter are flourishing: in 2018, Apple became the first company in history to reach $1 trillion in market capitalization. The Californian allegory of the two ecosystems tells us how successfully we are achieving the digital transition and how much we are failing the ecological transition. But can we reasonably argue that one is to the detriment of the other?

The beginning of the twenty-first century is characterized by a double temporal crisis. The first is due to the acceleration of the present under the effect of what is known as the "digital transition". The revolutionary nature of the latter is often exaggerated, but it is certainly deploying in both public space and private life incessant innovations that are altering human existence on a daily basis (attention capacity, quality of sleep) and deeply upsetting social organization (how to do politics without social networks today?). This acceleration of the present is reminiscent of the beginning of the twentieth century, when the conjunction of industrialization, globalization and urbanization agitated and disrupted Western societies. At that time, too, mental and social structures were shaken by the pace of technology.

The second-time crisis is rather reminiscent of the end of the 1960s: the future is obstructed by ecological crises, starting with climate change, in the same way as, at the height of the Cold War, the fear of the nuclear apocalypse blocked social horizons.

The particular difficulty of our time is that these two crises are occurring together; while they should reciprocally mitigate each other, they mutually aggravate each other. The headlong rush into innovation and frenetic change should open up new prospects for the future, as when, at the beginning of the twentieth century, we imagined the "city of the future". Similarly, uncertainty about the future generated by the threat of ecological crises should cause a social uprising here and now. Instead, we are stuck in transition: a promise of collective change that is not happening, all the while individual lives are perpetually accelerating.

The explanation that I propose is that the digital transition constitutes a growing obstacle to the ecological transition, so that the horizons of cooperation, as I have defined this notion (see Chapter 3) are doubly blurred by the crisis of the present and by that of the future. Contrary to the prevailing discourse which sees digital transition as a lever for ecological transition, there are multiple digital barriers to ecological transition (material, symbolic, psychological and social). Overcoming them means essentially decelerating digital transition to regain social continuity.

Let us first consider the damage caused by the digital transition in terms of cooperation. It is important to distance oneself from the soothing discourse that is too often heard, and according to which digital is merely a tool that can be what we decide to do with it. In reality, the digital transition is already here, and it is, on the contrary, what it makes of us. Because it is not only a question of a technological transition, but also of a transition of imaginations and practices.

The digital transition poses at least two major problems for cooperation as I have defined it in this book. To begin with, its founding mythology is that of the brilliant inventor connected to machines, but disconnected from society, this individual who, alone in his garage or in front of his laptop, is about to revolutionize the world. This figure of the nerd or geek has sympathetic aspects. The idea that offbeat or marginalized beings can become the "masters of cool" and thus take their revenge on a society whose cultural codes exclude them is something jubilant. However, the myth of the solitary programmer who "eats code" day and night and can do without cooperation is as false as it is pernicious.

It is false for two reasons: in the first place, because all the technological innovations of the digital transition have been made possible by the cooperation of teams mixing different specialties and whose joint work has enabled not only the invention itself, but above all innovations; secondly, because the great figures of Silicon Valley (Bill Gates, Steve Jobs, Mark Zuckerberg or even Elon Musk) are entrepreneurs much more than innovators (the case of Steve Jobs is undoubtedly the most instructive: he was never a developer and did not even know how to code).

This imaginary is also pernicious because many innovations of the digital transition aim, in line with nerd mythology, to delegate to machines the very function of social link (the algorithms that govern social networks or the deployment of artificial intelligence in customer services are good examples). Cooperation itself is delegated to machines, which are entrusted with the mission of socializing in place of humans. However, if we can to a certain extent mechanize individual intelligence (even in the complex case of the game of Go), it is impossible to mechanize collective intelligence.

Above all, the digital transition is an obstacle to cooperation as a quest for shared knowledge. Indeed, if cooperation is a human association whose goal is knowledge, then the illusion of a knowledge already constructed, ready for use and that it would suffice to locate rather than invent (as with ChatGPT) takes away its *raison d'être*. The new communication technologies induce collective unlearning by signaling the end of trial and error: the horizon of common knowledge is receding, because knowledge appears to be already there. Behind the curtain of hyper-collaboration, the digital economy is therefore the scene of an atrophy of cooperation.

But more importantly, digital acceleration is a war against time, especially free time. This war deserves to be put into perspective in the anthropological long term. It can be argued that humans have been engaged for thousands of years in a struggle to free up their time, in other words, in a war *for* time. Two stages marked this emancipation: the liberation of natural time and the liberation of social time.

Humans first freed themselves from physical time by managing to domesticate natural rhythms through agriculture and animal domestication, then by mastering increasingly complex sources of energy (from fire to solar panels) and technologies to save on labour. They then freed themselves from social time by obtaining that work occupies only a limited part of their existence, then that free time itself be compensated (this is the principle of paid leave). It would be wrong to think that this liberation of social time dates from the great popular struggles of the nineteenth century. In France, for example, it is a recent acquisition: in 1950, a Frenchman still devoted more than a thousand hours a year to work; this figure is 600 today (a French person devotes on average 17% of their time to work).

Nowadays, the most developed societies (not only the wealthiest in terms of income, but the most humanly advanced in terms of income, health and education) are also those in which free time is most abundant. Even within the OECD, which brings together the countries that are both the most developed on a human level and in which leisure time is the most extensive, we observe differences: on average, the more developed a country is on a human level, the more important the leisure-time recreation of its inhabitants.

When it comes to ecological crises, the gap between the progress of scientific knowledge and the inertia of political action has probably never been wider. The digital transition would have two major advantages in this respect: the de-materialization of economies and the acceleration of knowledge and catalysis of social organization.

Are we witnessing a digitally driven economic de-materialization? On the contrary, United Nations data (IRP 2017) show that in 2017 the world economy extracted three times more natural resources than in 1970. The beginning of the 2000s and the mid-2000s, when the digital transition was accelerating, marked an acceleration in extraction, neither a slowdown nor a stabilization (40 billion more tonnes were extracted between 2000 and 2017 compared to 20 billion more between 1970 and 2000). Even more striking, the relative decoupling of economic production and the consumption of natural resources observed throughout the twentieth century and until the early 2000s is then reversed. At a time when the digital revolution is supposed to make economies immaterial, humans today are extracting nearly 90 billion tonnes of material resources, i.e., more than three times the volume extracted 50 years ago (IRP 2017).

As such, humanity weighs almost nothing on Earth: when we estimate the respective biomass of the components of life on our planet, it appears that the total biomass weighs around 550 Gt C (gigatons of carbon), of which 450 Gt C (or 80%) are plants, 70 Gt C (or 15%) are bacteria and only 0.3% are animals. Within this last category, humans represent only 0.06 Gt C. In other words, the approximately 8 billion humans account, in volume, for only 0.01% of life on the globe (Bar-On et al. 2018). But this human population has been frantically producing for two centuries an ever-increasing "anthropic mass" to meet its needs: cement, sand, gravel, metals, asphalt, bitumen, cardboard, plastic. From 1900 to 2020, all of this human production amounted to 1,154 Gt, and by 2021 it exceeded total biomass (the weight of all life on Earth, including humans).[8] Cement alone weighs 9,000 times more than all of humanity, its mass having been multiplied by 50 since 1940. And each year we consume around 100 billion tons of natural resources to satisfy

[8] Elhacham et al. (2020).

our disproportionate and unequal needs, of which only 8 come from recycling.

But has the digital transition really accelerated the knowledge essential to the resolution of ecological crises? This is anything but obvious: nothing indicates that humanity has been tremendously more intelligent for twenty years. The growing intensity of ecological crises is rather a sign that we are losing the great race between understanding our living environment and destroying it. To take just the example of climate science, its incredibly robust broad lines were drawn up at the end of the 1980s, before the dawn of the digital transition. On the contrary, a recent study suggests that recent academic papers are becoming less and less "disruptive" all the while digital access is ubiquitous.[9]

Has the digital transition been a catalyst for social organization in favor of ecological transition? Here too, we have the right to be skeptical.

Ecological transition requires changing human attitudes and behaviors to consciously begin to preserve the biosphere rather than continuing to blindly destroy it. The digital transition contravenes this agenda in two ways.

First, because the digital transition is accelerating consumer desires, the limitless visual store that is Amazon is a striking illustration of this, and the absence of the slightest ecological price signal (the Amazon Prime system magically erasing even the delivery cost) further accentuates the unleashing of this desire.

Second, because it monopolizes individual attention and therefore limits the capacity for social cooperation. The metaphor of the sidewalk is quite enlightening in this respect: the eyes of city dwellers are riveted downwards (neither towards the sky, the direction of the imaginary, nor towards others, the direction of social cooperation). In doing so, the digital tools supposed to facilitate the orientation of individuals, complicate and slow down much more than they simplify and fluidify. Passers-by

[9] Park et al. (2023). The authors note: "Recent decades have witnessed exponential growth in the volume of new scientific and technological knowledge, thereby creating conditions that should be ripe for major advances. Yet contrary to this view, studies suggest that progress is slowing in several major fields. Here, we analyse these claims at scale across six decades, using data on 45 million papers and 3.9 million patents from six large-scale datasets, together with a new quantitative metric—the CD index12—that characterizes how papers and patents change networks of citations in science and technology. We find that papers and patents are increasingly less likely to break with the past in ways that push science and technology in new directions".

no longer look at each other, they no longer consider each other, are almost unaware of their surroundings and become obstacles for others. German sociologist Simmel explained well in his *Sociology of the Senses* the essential role of the eye as an organ of urban trust (it cannot see without being seen); it is this trust that is lost when the gaze is riveted on the digital device. The paradox is that while individuals are thus entertained (in the double sense of technological diversion and amusement), they are at the same time monitored (located, filmed, recorded) at each stage of their urban movements.

The digital transition also complicates and slows down cooperation over time: the permanent interruption of attention and the constant diversion make the continuity required by cooperation impossible. Technological intermittency is the enemy of social continuity and therefore of ecological transition.

What is more, the "low-tech" civic mobilizations of the 1960s and 1970s did infinitely more for the development of environmental policies than social networks (the same is true of the environmental revolts in the rural areas of contemporary China, which barely have access to the, in any case tightly controlled, internet). In addition, whether an individual is alone or accompanied by millions of supporters on social networks, they have the right to the same dignity in court.

So how to regain social continuity? It is not about reinventing human nature or forcing humans to cooperate, which would be a contradiction in terms. Rather, it is about working to reclaim our imaginations and reform our institutions. Imaginaries serve to give meaning to the past and to connect it to the present; institutions serve to plan for the future. Imaginaries shape values; institutions shape behavior. Focusing on imaginaries makes it possible to move away from a scientist vision of the social, according to which the members of a community act only out of interest in order to achieve collective efficiency.

In this spirit, I advocate for an "ecological luddism" on the model of social Luddism from the beginning of the nineteenth century: a conscious social movement to slow down the digital transition consisting in putting digital objects at a distance in space and time while domesticating them. Three major projects would become priorities under this agenda.

First, reinvent a genuine cooperative economy. In the Europe of the first half of the fifteenth century, the advent of urban civilization was accompanied by the emergence of an original way of conceiving social relations and economic life, rooted in humanist philosophy and largely forgotten afterwards. The "civil economy", which is based on institutions, laws and civic virtues, intends to make the market a place of human development aimed at the happiness of citizens through the social sharing of wealth and by relying on reciprocity in social relations. The authors who have rediscovered this current of thought identify a fundamental difference between civil economics and contemporary "economic science", which emerged from the neoclassical current in the nineteenth century and was based on individualism. As Luigino Bruni and Stefano Zamagni (both Italian scholars who revived the civil economy)[10] put it, the issue then was that the pursuit of private interest does not generate destructive antisocial dynamics and that the market, framed and nourished by other forms of civil and spiritual life, works in favor of society and not against it. Returning to the civil economy means prioritizing cooperation over collaboration in all social spaces, be it work, education or research. It also means rediscovering the richness and diversity of institutional forms that pre-existed the institutions of capitalism.

Second, decelerate cities. The "smart city" dystopia locks city dwellers into a political project that is exactly contrary to the vocation of anonymity and freedom that gave birth to the great cities of humanity. Urban spaces are by nature spaces of internal and external cooperation: it is to facilitate social cooperation that cities were invented and they were invented in a spirit of cooperation with their environment. The first cities were thus born not after, but with agriculture, at the turn of the Neolithic era, from 10,000 years before our own era.

Third, civilize "tech". Silicon Valley companies are based on an asymmetrical opacity: the absolute transparency of users contrasts with the almost total opacity of producers. These have established a formidable system of revealing consumer preferences (which exempts them from resorting to advertising to arouse desire) and social separatism (fiscal but also with regard to their management practices). A powerful protest movement is rising in North America on this subject, as evidenced by the resounding recent divorce of Amazon and New York City and the end of the "Google city" project in Toronto. While the public authorities are looking for new tax bases, they could consider a levy on digital companies aimed at financing the ecological transition (a tax on short-termism aimed at securing the long term). By the same token, children should be

[10] Bruni and Zamagni (2016).

protected, both from screen addiction and digital harm. Health professionals are alarmed in many countries by the sharp increase in cognitive disorders linked to digital uses. This is not an intergenerational problem but a health and political issue. By the same token, unnecessary nuisances such as the Internet of Things and autonomous cars whose primary purpose is permanent spying on users (within "surveillance capitalism") should be firmly regulated.

We are experiencing two transitions that are more and more opposed. The digital transition, described as inevitable and beneficial but which nothing in reality makes necessary, is accelerating every day before our eyes. The ecological transition, so often depicted as impossible and costly but on which the planet's hospitality for our species depends, is lagging considerably behind, for which we will pay dearly. The digital transition has an essential relationship with the ecological transition: by digitizing a world that we are destroying, it becomes the memory of our failure.

Mobilizing People and Activating States

Of all ecological crises, climate change, reputedly slow and invisible, is today the most apparent and the fastest. Everywhere on the planet, the observation is clear: climate change is no longer a threat but a scourge. An aggravating circumstance, the gap between the progress of scientific knowledge and the inertia of political action has probably never been so great, despite the progress made by the Paris Agreement negotiated, signed and ratified in record time. One explanation for this apparent paradox lies in the nature of the scientific knowledge that has been produced over the last 30 years: if what is commonly called "climate science" has made giant strides, advances in what could be called the "science of climate transition" are much more measured. Conversations about the urgency of the ecological transition indeed often end with a disillusioned call for "political will" (which is always lacking), but the dynamic of this will is rarely analyzed. In closing this chapter devoted to policy, I propose here to understand the ecological transition as a social dynamic that can be set in motion with, without and even against the state and should be considered as a two-way street. Social-ecological policy is about actioning people and auctioning states (which can in turn action people).

Let us start from the usual understanding of the ecological transition policy as the coherent implementation of a range of public policy instruments by the state with a view to achieving objectives defined, in the best of cases, by scientific expertise. A distinction is traditionally made between four types of these instruments:

1. The first instrument in the toolbox is awareness raising through information: the state here seeks to influence attitudes in order to modify behavior, in other words to alter the value system of individuals and groups to initiate actions deemed desirable. The essential lever here is not administrative constraint or economic incentive but the power of desire. One can think for instance about the comeback of bicycles in European cities under the effect of a change in the imagination of mobility partly supported by the public authorities, particularly in Paris since the mid-2000s.
2. The easiest public instrument to guide behaviors is administrative regulation, through which the state exercises its authority or even its strength over its constituents. Again in the European urban context, a good example is the gradual imposition of increasingly restrictive "low emission areas" in many urban centers such as Berlin. This instrument of the public arsenal of transition comprises two parts, the second of which is often neglected. It is not only a question of regulating, it is also necessary to check that the regulations are well respected (hence the notion of "command and control" policy).
3. In this state toolbox (the state being understood in the broad sense as a public power exercising itself at different levels of government), we find the instrument preferred by economists: market mechanisms including "price signals" emitted by Pigouvian taxation or pollution markets ("cap and trade") adorned with the supposed virtue of minimizing the cost of the transition by relying on the knowledge that the actors have of themselves. The success of the European carbon market (EU ETS) in reducing emissions on the continent (around 43% since its creation in 2005) has, for instance, convinced the European Commission and the Member States to extend it considerably (in particular to transport) from 2026 to become the central instrument of the European. Economists and those who listen to them tend to forget the extent to which market instruments are in fact regulatory instruments whose viability depends entirely on the

relevance of the rules that structure them (hence pollution markets that can be proving terribly inefficient and unfair).

4. Finally, public investment can be used to build or modernize infrastructures so as to accelerate ecological transition and foster sufficiency (cf. supra).

Three driving forces are supposed to ensure that the public impetus is effectively translated within civil society into a social dynamic setting the transition in motion: moral virtue, economic interest and the thirst for justice, the latter being undoubtedly the most mighty of the three.

But what about the converse proposition that ecological transition is just as much and perhaps even more a desire of civil society to overcome the resistance of the state to assume its role by making good use of all the instruments with which it is endowed to counter economic lobbies? What instruments does civil society have to make the ecological transition to the public authorities?

1. The first type of instrument is, as for the state, of an informational and educational nature: occupying the public space and in particular the media by making full use of the freedoms available to citizens in democracy and by mobilizing those that are still available in repressive regimes ("agenda power"). The first is to produce and disseminate scientific knowledge on environmental crises and their social consequences (in particular health) and counter-expertise to alert the community of citizens to the reality and seriousness of environmental changes, compelling public authorities, while they might not want it for fear of hurting vested interests (consider whistleblowers on toxic substances such as asbestos or phthalates or environmental justice activists in North and South America). This citizen mobilization also serves to denounce the public and private policies which hinder the transition and which are legion (subsidies for fossil fuels or phytosanitary products, regulations restricting the emergence of alternative models such as organic farming, investment projects in climate bombs such as new pipelines, lobbying, misinformation including in the academic field, etc.).
2. The second type of instrument consists in using the rule of law against the state by taking legal action, by filing appeals, by investing the bodies of public debate, by appealing before national, regional

and international courts to recall public powers to the law to which they are subject (for instance Urgenda Foundation *v.* The State of the Netherlands, June 24, 2015). But the arsenal of legal instruments is broad and goes beyond the direct questioning of the state because of its action or its inaction. It also includes legal recourse against companies, in particular in fossil-fuel and chemical industries, which have deliberately delayed or blurred awareness of ecological perils and whose acknowledgment of responsibility could lead to financial penalties which would be as many resources for the transition.

3. The third type of instrument covers the broad spectrum of economic instruments, many of which were used in the civil rights movement in the United States during the 1960s and 1970s: boycott/refusal to buy, purchasing power (commercial and financial, in particular placement of savings and rules governing investment), material impediment of economic activity by blocking, striking or sabotaging, development of alternative economic services (such as local currencies), etc.
4. Finally, and this is a major originality compared to the arsenal of the state, civil society has civic instruments which together form a range of means by which civil society assumes itself as a civic group and not an anonymous mass of individualized economic agents. In this category, we find, first and foremost, voting in all its forms (electoral and corporate ballots, participatory budgets, etc.), the organization of primaries and political movements and influence groups (unions, think tanks, etc.), invention of counter-models of social organization (ZAD, autonomous territories, etc.).

What, then, would be the springs that would put the state into transition under the impetus of civil society? We undoubtedly find the economic interest (determined actions can lead to a loss of fiscal resources and therefore hinder the public authorities' ability to act, particularly at the local level) but the most powerful is probably the state's fear losing its power over civil society and, more serious still, its loss of trust (trust which extends in the form of credibility in the international sphere). Thus, what would push the state to obey civil society would be the fear that the latter would no longer obey it.

In short: the ecological transition is not a one-way social dynamic but a two-way one and these two directions are equally important to consider;

the ecological transition is not a face to face between the state and civil society: it is a dynamic triangle that connects the part of civil society, now a majority, that wants the transition; the part, still a majority, of the business community which does not want it; and the state, whose powerful instruments must be activated by all possible means to finally make it happen.

References

Abrar, R. (2021), "Building the Transition Together: WEAll's Perspective on Creating a Wellbeing Economy". In: E. Laurent (ed), *The Well-Being Transition. Analysis and Policy*. London: Palgrave Macmillan.

Aldy, J. E., Kotchen, M. J., Stavins, R. N., and Stock, J. H. (2021), "Keep Climate Policy Focused on the Social Cost of Carbon", *Science*, vol. 373, no. 6557, pp. 850–852.

Bruni, L., and Zamagni, S. (2016), *Civil Economy: Another Idea of the Market*. Agenda Publishing.

Crutzen, P. J., and Stoermer E. F. (2000). « The "Anthropocene" », Global Change, IGBP Newsletter, no. 41, pp. 17–18.

EEA. (2021), *Growth Without Economic Growth. Briefing No. 28/2020*. https://doi.org/10.2800/492717

Elhacham, E., Ben-Uri, L., Grozovski, J. et al. (2020), "Global Human-Made Mass Exceeds All Living Biomass", *Nature*, vol. 588, pp. 442–444.

Gough, I. (2021), "Two Scenarios for Sustainable Welfare: New Ideas for an Eco-social Contract", *ETUI Research Paper—Working Paper, no. 12*.

Hale, T. (2021), "What We Learned from Tracking Every COVID Policy in the World", *The Conversation*, March 24, 2021. https://theconversation.com/what-we-learned-from-tracking-every-covid-policy-in-the-world-157721

Hickel J., Brockway P., Kallis G. et al. (2021), « Urgent need for post-growth climate mitigation scenarios », *Nature Energy*, vol. 6, pp. 766–768.

IPCC. (2021), "Climate Change 2021: The Physical Science Basis". In: V. Masson-Delmotte, P. Zhai, A. Pirani, S. L. Connors, C. Péan, S. Berger, N. Caud, Y. Chen, L. Goldfarb, M. I. Gomis, M. Huang, K. Leitzell, E. Lonnoy, J. B. R. Matthews, T. K. Maycock, T. Waterfield, O. Yelekçi, R. Yu, and B. Zhou (eds), *Contribution of Working Group I to the Sixth Assessment Report of the Intergovernmental Panel on Climate Change*. Cambridge University Press.

IPCC (2022), "Climate Change 2022: Mitigation of Climate Change. Contribution of Working Group III to the Sixth Assessment Report of the Intergovernmental Panel on Climate Change" [P. R. Shukla, J. Skea, R. Slade, A. Al Khourdajie, R. van Diemen, D. McCollum, M. Pathak, S. Some, P. Vyas,

R. Fradera, M. Belkacemi, A. Hasija, G. Lisboa, S. Luz, J. Malley, (eds.)]. Cambridge University Press.

IRP. (2017), "Assessing Global Resource Use. A Systems Approach to Resource Efficiency and Pollution Reduction. A Report of the International Resource Panel", United Nations Environment Programme. Nairobi.

Jacobson, M. Z. et al. (2022), "Low-cost Solutions to Global Warming, Air Pollution, and Energy Insecurity for 145 Countries", *Energy & Environmental Science*, vol. 8.

Janoo et al. (2021), Wellbeing Economy Policy Design Guide, Well-being economy Alliance.

Kikstra, J. S., Mastrucci A., Min, J., Riahi, K. (2021), "ND Rao Decent Living Gaps and Energy Needs Around the World", *Environmental Research Letters,* vol. 16, no. 9.

Laurent, E. (2021), Sortir de la croissance. Mode d'emploi, Les Liens qui libèrent, « Poche + », Paris.

Laurent, E. (2018), *Measuring Tomorrow. Accounting for Well-Being, Resilience, and Sustainability in the Twenty-First Century*. Princeton/Oxford: Princeton University Press.

Laurent, E. et al. (2022), "The Wellbeing Reflex: Facing Covid with a 21st Century Compass", WEAll Policy Brief.

Moore, J. W. (2013, May 13), "Anthropocene, Capitalocene, and the Myth of Industrialization, Part I", *World-Ecological Imaginations*.

Morelli, X., Gabet, S., Rieux, C., Bouscasse, H., Mathy, S., Slama, R. "Which Decreases in Air Pollution Should be Targeted to Bring Health and Economic Benefits and Improve Environmental Justice?" *Environment International*, vol. 129, pp. 538–550. https://doi.org/10.1016/j.envint.2019.04.07.

Oliu-Barton et al. (2021), SARS-CoV-2 elimination, not mitigation, creates best outcomes for health, the economy, and civil liberties. Lancet. 2021 Jun 12; no. 397 (10291), pp. 2234–2236.

Park, M., Leahey, E., and Funk, R. J. (2023), "Papers and Patents are Becoming Less Disruptive Over Time", *Nature*, vol. 613, pp. 138–144. https://doi.org/10.1038/s41586-022-05543-x

Parrique, T. "The Political Economy of Degrowth. Economics and Finance", Thesis, Stockholm University, 2019/Université Clermont Auvergne, 2017–2020.

Philippe, D. and Marques, C. (2021), The Zero Covid strategy protects people and economies more effectively, Institut économique Molinari. France.

Pörtner, H. O. et al. (2021), *IPBES-IPCC Co-sponsored Workshop Report on Biodiversity and Climate Change*. IPBES and IPCC.

Prados de la Escosura, L. (2015), World Human Development: 1870–2007. *Review of Income and Wealth*, vol. 61, pp. 220–247.

Ramsey, F. P. (1928), "A Mathematical Theory of Saving", *The Economic Journal*, vol. 38, no. 152, pp. 543–559.

Raworth, K. (2012), "A Safe and Just Space for Humanity. Can We Live Within the Doughnut?", Oxfam Discussion Paper, February 2012.

Stern, N., and Stiglitz, J. E. (2021), "The Social Cost of Carbon, Risk, Distribution, Market Failures: An Alternative Approach", NBER Working Paper, no. 28472.

Vogel, J. (2021), "Socio-economic Conditions for Satisfying Human Needs at Low Energy Use: An International Analysis of Social Provisioning", *Global Environmental Change*, vol. 69.

CHAPTER 5

Narrative: Reimagining Economics

Stories are the roadmaps for human action, especially collective action. Jean-Paul Sartre (1938) wrote on this subject: "A man is always a storyteller, he lives surrounded by his stories and the stories of others, he sees everything that happens to him through them; and he tries to live his life as if he were telling it".

Anchoring social-ecological transition in a narrative therefore does not consist in making things up but in recounting human experiences and recognizing that any collective transition is part of a triptych associating ideas, institutions and interests, components of human transitions linked together by a social catalysis: ideas irrigate institutions that shape interests. In this catalysis, which aims to build institutions capable of lasting change in behavior and attitudes, narratives and their contradictions play a key role.

For instance, Biblical stories (and in particular the tension, in Genesis, between the story of human domination over the biosphere and the story of human responsibility with regard to the biosphere) and mythological stories (such as the myth of Prometheus, which tells the story of an emancipation through technology, but also an emancipation of technology).

É. Laurent, *Toward Social-Ecological Well-Being*, Palgrave Studies in Environmental Sustainability, https://doi.org/10.1007/978-3-031-38989-4_5

In closing this book, four contemporary narratives seem especially relevant to understand obstacles and leverages toward social-ecological well-being as I have attempted to define it.

The Doom Narrative

Fatalism regarding the pursuit of the human adventure on Earth dates from the very beginning of this adventure and since then there have been countless apocalypses that have been narrowly avoided. Ecological crises lend themselves particularly well to this slope of "doomism" whose principle goes back to the Greek tragedy. Ecological fatalism translates into two partly competing narratives: humans can no longer do anything about their fate, now determined by forces beyond their control; human nature is stronger than anything and condemns us to collective ruin.

But contemporary ecological crises are of human origin, so that human responsibility is decisive in their mitigation (even if it is true that "feedback loops" can cause chain reactions in the biosphere). As for the fatalism of the "tragedy of the commons", it is contradicted, as we have seen in detail, by the work of Ostrom which has irrigated this book and its social-ecological approach. The path toward social-ecological well-being is collective intelligence.

The Tech Narrative

An inverted mirror of the previous narrative, technologism promises to solve everything without changing anything: without altering either attitudes or behavior of humans, "disruptive" technological inventions would make it possible to mitigate our ecological crises in the coming years (this is notably the promise of "geo-engineering" in all its forms but also artificial intelligence).

Why go to the trouble of changing our economic systems from top to bottom as I have proposed and completely rethinking our individual and collective behavior if we are on the eve of technological advances that could save us effortlessly? And what better way than the stimulation of private finance to accelerate their advent? Isn't that how humanity was able to cope with Covid?

But messenger RNA vaccines, which have undeniably provided the best protection against Covid for the small part of the world's population that has been able to fully benefit from them, owe almost nothing to start-ups backed by business angels and other venture capitalists. They are the result of half a century of collective intelligence constantly supported by public funding (Dolgin 2021).[1]

The tech narrative, which is intended to be anticipatory, is in reality fundamentally wait and see, betting uncertainly on unproven technologies, such as carbon capture, "albedo enhancement", "space reflectors" or "stratospheric aerosols". However, the ecological transition, starting with the energy-climate transition and the agricultural transition, is not currently hampered by the absence of technological means. As Mark Jacobson writes, "no miracles needed" (Jacobson 2023): we have for instance all the technologies we need to implement 100% renewable scenarios for climate-energy strategy (this is even truer for the development of agro-ecology, the technologies of which have been known and mastered for millennia). Conversely, the transition is also a matter of "exnovation", that is to say the exit from certain innovations that have become destructive on the environmental level: how to put an end to the thermal vehicle and even more to the individual car that occupies our cities as well as our imaginations?

More fundamentally, the narrative of technologism misses the point: what we really lack is social innovation. From this point of view, the re-development of the bicycle in many of the most advanced cities in social-ecological transition such as Copenhagen is rich in lessons: it is firstly because it has imposed itself culturally that the public authorities have had to develop cities to welcome it. Nor is it a breakthrough technology: invented in the nineteenth century in its modern form, it was present in cities all over the world at the beginning of the twentieth century before being driven out by cars. Finally, the more technological the bike is (in particular when equipped with electric batteries), the less important are its health and environmental benefits.

[1] The discovery of messenger RNA dates back to the early 1960s, but it was not until the early 1990s that the first vaccines based on this technology were tested. Although BioNTech and Moderna were created in 2008 and 2010 respectively, it was the decision taken in 2012 by the American government to massively fund research in this field which accelerated progress, which until then had been modest, leading in 2015 to the first clinical test of a messenger RNA vaccine.

The Economic Narrative

Economism promises to make the story of technologism fully operational: the system of economic valuation through the use of prices and the market is supposed to mechanically bring about the technologies that consume the least natural resources and are the least destructive of the biosphere. This story collides head on with a reality that is hard to contest: the period that began at the beginning of the 1980s, whose neoliberal label is widely accepted, was marked by strong and generalized liberalization, but it was also a period of unprecedented acceleration of ecological crises.

The wanderings of the energy markets clearly embody the limits of the narrative of economism: while they have been liberalized and globalized, their price signals scramble and complicate the low-carbon energy transition, while renewable energies have been able to develop through public intervention.

This false narrative of economism is grounded in an academic field which, contrary to appearances, is in serious crisis.

For starters, economists themselves are expressing genuine unease about the state of their discipline. A recent study published by the IZA (IZA 2021), a reputable German research institute, clearly shows that a clear majority of the researchers surveyed express intellectual frustration with the research topics most valued by the profession and the objectives of their own research. Respondents also believe that economic research should be reformed to become more policy relevant, more multidisciplinary, and address more varied and innovative topics. The authors of the study conclude that the discipline of economics as it has become does not do justice to the aspirations of economists.

Then a lack of constructive academic dialogue among economists has emerged: according to the calculations of the economist of the London School of Economics Joe Francis (Francis 2014), the number of economic works published in response to others to contest or qualify their argument gradually increased in the first half of the twentieth century until it represented a quarter of academic publications in 1970. The economists of that time knew each other, spoke to each other, refuted each other. Today, only 1% of the articles published in the five most consulted economic journals (so-called top 5) are devoted to criticizing the results of other work. The academic debate between economists has weakened compared to what it was during the twentieth century.

Third, a serious problem in the teaching of economics was brought to light at least twenty years ago, when student groups in North America and Europe denounced the lack of intellectual pluralism in the courses of economy and their neoliberal orientation. This dissatisfaction remains strong, as evidenced by the vigor of the student movement "Rethinking Economics" which published a new Manifesto at the beginning of 2022 which proposes to "reconquer the economy". The students who lead this movement write on this subject:

> *Human economic activity is unsustainable, fuelling global heating and destruction of the living world, which will make large parts of the earth uninhabitable, and already causes untold damage to people in countries with the least historical responsibility for causing this crisis. Until we fundamentally reorganise local, national and global economies, we will not be able to address the interlinked challenges of systemic racism, gender-based inequality, the Covid-19 pandemic and the nature and climate emergency.*
>
> *Society entrusts the discipline of economics with the tasks of understanding our economies and educating the economists who go on to guide government and business. But economics has failed to grapple seriously with the crises we are facing, so we are left without access to the knowledge, skills or tools we need to build better futures. Across the world, economics students are coming together under the banner of the student movement, Rethinking Economics, to create a better economics—one which can help to create a world where all our children can flourish regardless of their gender, background or birthplace.* (Ambler et al. 2022)

Finally, economists are at odds with the world they inhabit.[2] The vast majority of economists ignore environmental issues, in the double sense of ignorance and indifference.[3] When they do care, it is generally to

[2] "Economists" have existed as such since the 1750s, when the physiocrats imposed their analyses and their concepts on court conversations in France, to the point that their detractors began to see in their current of thought a "cult".

[3] To try to shed light on this question, we can first refer to the study published by Andrew Oswald and Nicholas Stern aimed at evaluating the place of environmental issues in academic publications in economics. Out of 77,000 articles published in the ten most influential journals in the discipline, exactly 57 were devoted to climate change, i.e., less than 0.1% (Oswald and Stern 2019). According to another accounting, we can show that, out of 44,000 articles published since 2000 in 50 reference journals, eleven have been devoted to the decline of biodiversity, again around 0.1% (Goodall and Oswald 2019). We can add to these bibliometric data which relate to the volume of publications the issue of their recognition. We then see that, of the 20 articles considered in 2011 as

minimize its impact and to propose, to alleviate contemporary ecological crises, remedies that aggravate them, such as the acceleration of economic growth or the monetization of ecosystem services.

Boyce (2020) thus shows that the increase in global average temperature that would accompany the "optimal" carbon price recommended by the 2018 recipient of the Nobel Prize in economics William Nordhaus (within his DICE model) is 3.5 °C by 2100 and continuing to rise thereafter. The DICE model therefore recommends a temperature rise twice as high as that of the scientific consensus patiently developed over three decades—on the basis of a fragile methodology, explicitly called into question by the IPCC.[4]

There is no doubt that there are thousands of economists around the world, who are genuinely concerned about ecological issues and work usefully to understand and mitigate them (starting with the members of the Intergovernmental Group of Experts on Climate Change—IPCC—and the Intergovernmental Science-Policy Platform on Biodiversity and Ecosystem Services—IPBES). But there is also no doubt that they represent a small minority (close to 5%) in the vast field of economic research and decision making.

This lack of interest in environmental issues within the economic discipline is all the more surprising in that it was affirmed and prolonged when empirical economics gradually began to prevail over theoretical economics (three-quarters of the articles currently published are considered empirical compared to 48% at the beginning of the 1960s, while pure theoretical articles only represent 19% of publications compared to 50% at the beginning of the 1960s (Hammermesh 2013).

This lack of interest in environmental issues is all the more detrimental since, as this book has attempted to show, the ecological transition is now a social science issue: the hard sciences have largely worked to

the most important in a century of existence of the *American Economic Review* (Arrow et al. 2011) by eminent representatives of the discipline, none deals with environmental issues. Of the 100 most cited economists listed by the Ideas/Repec website, not one is an environmental economist. Of the 100 most cited works listed by the Ideas/Repec site, not one deals with environmental economics. Of the 70 most cited articles in the five most influential academic journals in economics during the period 1991–2015 (i.e., 1% of articles), none deals with environmental issues (Linnemer and Visser, 2016).

[4] Ironically, Nordhaus was awarded his economics prize on the day of the release of the IPCC report showing how much humanity would benefit from limiting global warming to 1.5°C rather than 2°C.

reveal the extent and urgency of the crises ecological. We now need to change attitudes and behaviors to prevent human well-being from self-destructing over the coming decades. In other words, it is the social sciences, including economics, that hold the key to the problems that the hard sciences have revealed.

Twentieth-century economics, which is still professed by the overwhelming majority of professional economists and practiced by governments around the world, crystallized between 1934 and 1936, under the cross-influence of Simon Kuznets and John Maynard Keynes and on the basis of neoclassical economics. While in the aftermath of the Great Depression Kuznets invented the reference indicator supposed to measure collective wealth, the gross domestic product (GDP), Keynes designed the instrument likely to make it grow: macroeconomic policy. Shortly after the Bretton Woods conference, in November 1944, the second Beveridge report ("Full employment in a free society") appeared, linking economic growth and full employment. Growth, macroeconomic policy and full employment: three concepts brought to light in a decade from 1934 to 1944, and which were to form the triptych of social progress for the following 80 years, until now.

The twenty-first century probably began on April 7, 2020 when 4 billion humans were locked down by half of the governments of the planet to protect them from an unknown and deadly virus generated, it is now a virtual certainty, by the destruction of ecosystems and the commodification of biodiversity. At the time of writing, nearly 20 million human lives have been lost as a result of the COVID-19 pandemic and the virus has not struck at random: it has taken away the most vulnerable and weakened the most fragile.

The economy in the twenty-first century must therefore be an embedded economy, limited upstream by biophysics, with, as a frontier discipline, ecological economics (which studies the flow of materials, waste, energy, biodiversity, ecosystems, etc.) and bounded downstream by social justice, with, as a frontier discipline, political economy (which highlights social inequalities and measures the quality of political institutions). And it is an economy of essential well-being, which articulates universal human needs with planetary ecological constraints by projecting them over time, which brings us to the social-ecological narrative.

The Social-Ecological Well-Being Narrative

The social-ecological well-being narrative is, at its core, a return to the essentials of humanity: social ties and natural ties, intertwined. It is by renewing our natural ties that we will reweave our social ties: the power of our social cooperation put at the service of the perpetuation of life in a social-ecological loop whose knots are health and social ties; the war against the biosphere stopped by social peace; and life preserved by justice, justice saved by life.

This story is as old as the world, from the Greek *polis* ecology of gods, land, and people to the medieval European Great Chain of Being, the multi-species genealogies of Maori tribes or the contemporary protection of the Amazon by Indigenous communities. And it has contemporary meaning: according to the IPBES (IPBES 2022), we can distinguish at least four possible modalities of articulation between social and natural systems, highlighting the richness of these relationships beyond instrumental exploitation: living from nature emphasizes the ability of nature to provide resources such as food and material goods; living with nature emphasizes the life of non-human living beings, for example the intrinsic right of a fish to live freely in a river; living in nature refers to the importance of nature as a framework contributing to forging a sense of belonging and the identity of people; and living like nature illustrates the physical, mental and spiritual connection of human beings with nature.

In his famous triptych *The Garden of Earthly Delights* completed in 1500, Hieronymus Bosch shows a tragic representation of the destiny of humanity that resonates strongly in our own era of ecological crises. In the first panel, Adam and Eve enjoy themselves in the Garden of Eden; in the central panel, already too many humans occupy all the space and consume resources that are obviously too scarce; in the last panel, humans are subjected to horrible tortures in the Underworld. This is a striking vision of the unreason of humanity, to whom a hospitable planet has been offered and which, for lack of moderation and temperance (of sufficiency, one would say today), ruins it at its own costs. But this is a moralistic narrative that articulates abundance with excess and ultimately punishment.

Research in social science is centered not on the validity of the moral values of individuals but on the robustness of institutions, the best of which are supposed to resist individualistic excesses. The revolution of the commons initiated by Ostrom consists precisely in denying the moral fatalism of the ecological crisis by meticulously bringing to light the institutions of the social-ecological transition. At the end of this patient analysis, hope prevails.

References

Ambler, L., Earle, J., and Scott, N. (2022), *Reclaiming Economics for Future Generations*. Manchester University Press. http://www.jstor.org/stable/j.ctv29mvt6p.

Arrow Kenneth, J., Douglas Bernheim, B., Feldstein, M. S., McFadden, D. L., Poterba, J. M., and Solow, R. M. (2011), "100 Years of the American Economic Review: The Top 20 Articles", *American Economic Review*, vol. 101, no. 1, pp. 1–8.

Boyce, J. (2020), « Les dividendes du carbone. Le cas des États-Unis », *Revue de l'OFCE*, vol. 165, no. 1, pp. 97–115.

Dolgin, E. (2021), "The Tangled History of mRNA Vaccines", *Nature*, vol. 597, no. 7876, pp. 318–324.

IPBES (2022), "Summary for Policymakers of the Methodological Assessment of the Diverse Values and Valuation of Nature of the Intergovernmental Science-Policy Platform on Biodiversity and Ecosystem Services".

Francis, J. (2014). "The Rise and Fall of Debate in Economics", *History, Numbers and Some Theory*. August 29.

Goodall, A. H., and Oswald, A. J. (2019), "Researchers Obsessed with FT Journals List are Failing to Tackle Today's Problems", *Financial Times*, May 8.

Hamermesh, D. S. (2013), "Six Decades of Top Economics Publishing: Who and How?" *Journal of Economic Literature*, vol. 51, no. 1, pp. 162–172.

IZA. (2021), "What's Worth Knowing? Economists' Opinions About Economics", IZA Discussion Paper, no. 14527, July.

Jacobson, M. Z. (2023), *No Miracles Needed: How Today's Technology Can Save Our Climate and Clean Our Air*. Cambridge University Press.

Linnemer, L., and Visser, M. (2016), "The Most Cited Articles from the Top-5 Journals (1991–2015)", CESifo Working Paper Series, no. 5999, July 7.

Oswald, A. J., and Stern, N. (2019), "Why Does the Economics of Climate Change Matter so much, and Why has the Engagement of Economists been so Weak?", *Royal Economic Society Newsletter*, October.

Sartre, J.-P. (1938), *La Nausée Collection Blanche*. Gallimard.

Index

A

Agriculture, 4–6, 20, 87, 125, 129
Air pollution, 22, 24, 27, 35, 75, 80, 81, 113, 119, 120, 122, 123
Anthropocene, 21, 100
Aristotle, 29, 106

B

Beyond GDP, 103
Biodiversity, 1, 11, 20, 21, 23, 31, 33, 59, 62, 63, 68, 72, 90–94, 100, 105, 113, 123, 141, 143
Biosphere, 1, 8, 11, 13, 21, 31, 33, 34, 36, 39, 56, 58, 59, 67, 68, 92, 100, 103, 127, 137, 138, 140, 144
Boyce, James, 35, 91, 142
Brundtland Report, 7, 32

C

Capitalism, 100, 129
Carbon price, 142
Carbon tax, 38, 78, 82
Climate adaptation, 21
Climate change, 22, 23, 26, 27, 32, 33, 35, 77, 84, 109, 113, 118, 122, 123, 130
Climate policy, 84
Climate science, 127, 130
Co-benefits, 27, 64, 83, 111, 115, 118–123
Commons, 42, 86, 87, 90, 93, 144
Cooperation, 11–13, 38–43, 45–48, 51, 55, 57–60, 85, 86, 88–90, 95, 106, 117, 122–125, 127–129, 144
Cost–benefit analysis (CBA), 111–113, 115
COVID-19, 11, 18–20, 26–29, 49, 62, 64, 74, 75, 78, 111, 115–118, 120, 138, 143

D

Democracy, 48, 66, 86, 109, 132

É. Laurent, *Toward Social-Ecological Well-Being*, Palgrave Studies in Environmental Sustainability, https://doi.org/10.1007/978-3-031-38989-4

Digital tools, 11, 50, 61, 128
Digital transition, 12, 47, 50, 123–128, 130
Disability-adjusted life year (DALY), 24, 69, 120
Donut Economy, 55–57, 102–104

E

Ecological economics, 143
Ecosystems, 1, 8, 11, 20, 21, 23, 31, 56, 59, 62–64, 68, 72, 77, 91–94, 100, 105, 113, 123, 143
Ehrlich, Paul, 5, 6
Energy, 1, 33, 36, 43, 51, 56, 58, 59, 63, 79, 82, 94, 103, 104, 107–109, 111, 119, 122, 125, 139, 140, 143
Energy efficiency, 107
Environmental health, 22, 27, 62
Environmental inequality, 79–81
Environmental justice, 92, 121, 132
Environmental risk factors, 27
Environmental taxation, 82, 83
Equity, 11, 28, 107
European Environment Agency (EEA), 8, 22, 83, 105
European Union (EU), 24, 27, 70, 74, 77, 120
Exposure, 22, 24, 27, 28, 76, 80, 81, 119, 120
Externalities, 90
Extraction, 21, 58, 94, 126

F

Finland, 70, 105, 116–118
Food, 1–3, 11, 21, 28, 63, 76, 82, 108, 120, 122, 144
Fuel poverty, 79, 82

G

Green Deal, 77, 105
Gross domestic product (GDP), 5, 8, 9, 18, 90, 100–104, 115, 117, 122, 143
Growth, 2, 3, 5, 6, 8, 9, 13, 18, 21, 26, 33, 35, 36, 66, 67, 74, 78, 82, 90, 93, 99–105, 114, 122, 127, 142, 143

H

Happiness, 2, 11, 49, 63, 68, 106, 117, 129
Hardin, Garett, 40–42, 86, 89
Health inequalities, 28
Heatwaves, 26, 65, 75
Hotelling, 7

I

Intergovernmental Panel on Climate Change (IPCC), 8, 9, 21, 23, 31, 104, 107, 108, 119, 122, 142
Intergovernmental Science-Policy Platform on Biodiversity and Ecosystem Services (IPBES), 8, 21, 31, 72, 91, 94, 104, 142, 144

J

Just transition, 29, 60, 76–80, 82–86

K

Keynes, John Maynard, 101, 143

L

Life expectancy, 11, 17, 18, 24–27, 67, 69, 72, 74, 75, 106, 116, 119, 120

M

Malthus, Thomas, 2, 3, 5, 11, 12

Meadows, Denis, 6
Meadows, Donella, 6
Mental health, 22, 28, 59, 61, 63, 67, 72, 75, 110, 117
Mill, John Stuart, 6
Morbidity, 21, 65, 69, 70, 74, 75
Mortality, 20, 21, 25, 26, 49, 63, 69, 70, 72, 74, 75, 119

N
Narratives, 13, 55, 116, 137, 138
Nature's contributions, 21
Négawatt, 107
New Zealand, 105, 109, 116–118
Nordhaus, William, 142

O
Olson, Mancur, 40–42, 87
One Health, 62, 64
Organization for Economic Co-operation and Development (OECD), 18, 19, 24, 25, 31, 36, 50, 70, 74, 75, 84, 102, 105, 113, 115–117, 126
Ostrom, Elinor, 42, 86–90, 94, 138, 144

P
Physiocrats, 141
Piketty, Thomas, 30
Planetary boundaries, 8, 29, 33, 34, 56, 57, 103, 107
Planetary health, 10, 11, 17, 19, 21, 55, 58, 60, 61
Pollution, 6, 8, 19, 22, 23, 27, 32, 64, 74, 77, 80, 88, 113, 120, 122, 131, 132
Poverty, 34–36, 67, 108–110

R
Ramsey, Frank, 113
Raworth, Kate, 55, 103, 104
Regeneration, 61, 86, 90, 94
Resilience, 28, 29, 37, 59, 68, 72, 101
Ricardo, David, 3–7, 11, 12

S
Sen, Amartya, 63, 106, 109
Sensitivity, 22, 28, 80, 81
Smith, Adam, 41–45, 101, 102
Social discount rate, 113, 114
Social-ecological loop, 55, 56, 58, 59, 144
Social-ecological policy, 83, 130
Social ecology, 32
Social health, 64, 72
Social isolation, 48–50, 59, 64, 65, 72, 75, 122
Social justice, 12, 27, 58, 78, 80, 81, 85, 86, 143
Social links, 61, 91, 125
Social relations, 64, 65, 91, 129
Steady state, 4
Sufficiency, 29, 74, 106–109, 132
Sustainability, 1–9, 12, 13, 17, 33, 38, 51, 55, 58, 101, 112, 122
Sustainable development, 7, 56, 57

T
Transition, 48, 51, 55, 57–59, 69, 77–79, 83, 85, 86, 110, 111, 118, 121–124, 127, 128, 130–134, 137, 139, 140, 142, 145
Trust, 38, 39, 42, 47, 66, 85, 88, 89, 128, 133

V
Valuation of natural resources, 91

Vulnerability, 21, 22, 26, 28, 33, 63, 70, 72, 74, 79, 81, 93, 120

W

Waste, 35, 56, 112, 143
Water, 1, 22, 27, 28, 63, 86, 87, 103, 107
Wealth, 12, 13, 30, 34, 35, 58, 76, 82, 129, 143
Well-being, 6–9, 11–13, 18, 21, 22, 27, 29, 31, 32, 44, 55, 57, 58, 61, 63, 65, 68, 69, 75, 77–79, 83, 101, 103–107, 109–111, 113–120, 122, 138, 143, 144
Well-being budget, 104, 109, 110
Well-being economy, 64, 102–104, 110, 111
Well-being Economy Alliance (WEAll), 103, 110
World Health Organization (WHO), 22, 24, 27, 61–64, 68–70, 72

Z

Zoonoses, 20